TECHNICAL CHANGE, RELATIVE PRICES, AND ENVIRONMENTAL RESOURCE EVALUATION

Technical Change, Relative Prices, and Environmental Resource Evaluation

V. KERRY SMITH

Published by Resources for the Future, Inc.
Distributed by The Johns Hopkins University Press
Baltimore and London

RESOURCES FOR THE FUTURE, INC.
1755 Massachusetts Avenue, N.W., Washington, D.C. 20036

Resources for the Future is a nonprofit corporation for research and education in the development, conservation, and use of natural resources and the improvement of the quality of the environment. It was established in 1952 with the cooperation of the Ford Foundation. Part of the work of Resources for the Future is carried out by its resident staff; part is supported by grants to universities and other nonprofit organizations. Unless otherwise stated, interpretations and conclusions in RFF publications are those of the authors; the organization takes responsibility for the selection of significant subjects for study, the competence of the researchers, and their freedom of inquiry.

This book is one of RFF's studies on natural environments, directed by John V. Krutilla. V. Kerry Smith, formerly a research associate at RFF, is associate professor of economics at the State University of New York at Binghamton. The charts were prepared by Frank and Clare Ford. The book was edited by Margaret Ingram.

RFF editors: Mark Reinsberg, Joan R. Tron, Ruth B. Haas, Margaret Ingram.

Manufactured in the United States of America

The Johns Hopkins University Press, Baltimore, Maryland 21218
The Johns Hopkins University Press Ltd., London

Library of Congress Catalog Card Number 74-6840
ISBN 0-8018-1626-2

Library of Congress Cataloging in Publication data will be found on the last printed page of this book.

Contents

Foreword

During the past decade, legislation intended to protect and preserve specimens of the natural environment has imposed a substantial burden of analysis and evaluation on land management agencies. The Wilderness Act of 1964, for example, directed public land management agencies to review all unroaded areas of 5,000 acres or larger within their jurisdictions for possible inclusion within the national wilderness preservation system. The Wild and Scenic Rivers Act of 1968 similarly obligated public land managers to evaluate the relative benefits from alternative uses to which wild land and water areas may be put. Literally many hundreds of tracts of land will require such evaluation over the span of a relatively few years.

The problem of weighing the relative value of alternative, incompatible uses of wild lands and rivers was a matter of some expressed frustration by the Public Land Law Review Commission, whose June 1970 report complained of the absence of information in public land management agency files specifically relevant to such analysis. There are perhaps two difficulties here. Since each parcel of land will differ in important respects from any other, the relative value of alternative, incompatible uses will not be uniform for all tracts of land. At the same time, a careful, independent analysis of each specific tract involves analytical resources quite outside the ability of land management agencies, as presently staffed, to mobilize.

Apart from the problem of mobilizing resources, a special problem of methodology may arise when evaluating the benefits from different purposes to which a given tract of land or site may be put. The influence that technology has had on the relative annual benefits of the service flow of alternative uses must be taken into account if there is reason to suspect that it causes systematic change over time in those relative benefits. In traditional benefit-cost analysis, the convention of assuming that price levels remain constant has been adopted largely to avoid confusing simply pecuniary with real output effects in evaluation. Not until rather recently, and in relation to the evaluation of environmental modifications affecting wild and scenic lands and rivers, has the potential significance of systematic changes in relative benefits induced by differential incidence of technological change been perceived and incorporated into analysis.

In this study Kerry Smith investigates the conditions that would need to obtain on both the supply and demand side for goods and services of different characteristics in order for them to experience relative price appreciation. Some goods or services may be susceptible to fabrication or production, on the one hand, admitting of gains in productive efficiency through technological advance. Others, such as amenity services of the natural environment, are essentially not producible by man and thus cannot be augmented by advances in productive efficiency. Another class may not be either competitive or complementary in any particular instance, and, while producible, will combine in various proportions with one or another of the former in satisfying consumer wants. The relationship between these goods and the differential implications of technological change among them, giving rise to relative price appreciation, is the subject of Smith's investigation.

Smith demonstrates that under rather general conditions, intuitively satisfying, the rise in the relative prices of a class of goods or services of interest to us can be linked directly to the rate of general technological progress. Given this demonstration, resource development and environmental preservation alternatives can be evaluated with somewhat greater sophistication than in the past. The errors in evaluation that arise from failure to incorporate changes in the relative values of alternative uses are avoided; at the same time, analyses are not so complicated that meaningful evaluation by land management agencies would be impossible with the resources presently at their disposal.

Accordingly, while the Smith study is not intended for the perusal and direct use of land management agencies and their personnel, being directed instead toward economic theoreticians, it does provide the fundamental underpinning for simpler and more accurate evaluative techniques in land use planning.

John V. Krutilla, Director
Natural Environments Program
Resources for the Future, Inc.

Acknowledgments

Initiated in 1970 with a grant from Resources for the Future to me at Bowling Green State University and largely completed during my tenure at RFF, this study aims to provide a simple analytical framework for linking relative price behavior to the forces of technical change and intertemporal externalities, since they have important implications for the allocation of natural environments. John Krutilla, Director of Resources for the Future's Natural Environments Program, originally proposed that I focus upon these problems, and his stimulation and comments throughout have benefited the study immeasurably. Charles J. Cicchetti also contributed to the work in the formative stages and throughout the course of my research.

My sincere thanks are also due R. Talbot Page, who reviewed the entire manuscript and made numerous constructive suggestions. While he may not completely agree with the final product, Toby's perceptive criticisms improved the manuscript in all respects. Hirofumi Uzawa also took time from his busy schedule to review the book, and I am most grateful.

Helpful remarks on the literature review in chapter 2 came from Murray Brown from the State University of New York at Buffalo, William Fellner of Yale University, Thomas Hall of Phillips Petroleum Corporation, and Vernon Ruttan from the University of Minnesota.

Extensive comments from Clifford Russell and Karl-Göran Mäler on the general methodology helped me to avoid errors and opened up additional insights. Charles Metcalf and Robert Haveman also offered suggestions during early drafts of the models.

Seminars at the University of Michigan School of Natural Resources and the Natural Environments Workshops sponsored by RFF at the University of Montana School of Forestry and at Bowling Green State University provided useful ideas.

Others of my colleagues at Resources for the Future offered suggestions along the way; thanks especially are due Anthony C. Fisher and Allen V. Kneese.

I am indebted to Margaret Ingram for her careful and constructive editing. My special thanks goes to Rita Gromacki, whose typing skill through a number of drafts and endless symbols eased my work greatly. Thanks are also due Nettie Rathje, Irma Sedor, and Sandra Soltis for their efficient handling of the revisions to the manuscript.

Finally, I am most grateful to my wife, Pauline, whose patience and encouragement have been essential to all that I do. All shortcomings in the design and presentation of this work are, of course, my responsibility.

V. Kerry Smith
Binghamton, New York
October, 1973

chapter one Introduction

Natural environments are gifts of nature, dependent upon accidents of biological evolution, geomorphology, and ecological succession. These environments, the source of all goods and services the community requires, cannot be produced by man. Advances in production technology tend to augment the supply of man-made goods and services but do not increase our ability to produce the amenity services furnished by such environments in their natural state.[1] Therefore, deciding whether to develop a natural environment (i.e., to extract or somehow alter its resources) or to preserve it relatively untouched requires careful analysis if the correct decisions are to be made. Three recent legislative actions, the Wilderness Act of 1964, the Wild and Scenic Rivers Act of 1968, and the National Environmental Policy Act of 1969, have drawn attention

[1] I refer here to a particular class of amenity services, namely those associated with the recreational use of natural environments, for which there are few acceptable substitutes. Further, although improvements in transportation technology will open up certain natural environments previously inaccessible, this alteration does not represent a global increase in the supply of amenity services. Moreover, if each area is unique, no two will be perfect substitutes.

to the problems associated with allocating such resources to their most appropriate uses.

Extractive technology has advanced to the point where the supply of most of the materials provided (e.g., minerals) has been expanding at constant or falling relative supply prices (see Barnett and Morse 1963). On the other hand, as Fisher, Krutilla, and Cicchetti (1972) have recently noted in their evaluation of the Hells Canyon reach of the Snake River in Idaho and Oregon, the relative value over time of the amenity services provided by a scenic area in its preserved state is likely to appreciate. Clearly such an increase in relative value must be taken into account in the analysis of preservation versus development. There may be as many as 2,000 tracts of public lands for which similar decisions will need to be made in the next ten to fifteen years. The methodology outlined by Fisher, Krutilla, and Cicchetti is well suited to addressing the economic issues associated with these decisions. However, if that methodology is to be applied in such a diverse array of cases, the nature of the relationship between the rate of change of the relative value of amenity services in response to technological change and alternative demand and supply configurations must be more precisely defined. The purpose of this monograph is to develop models capable of explicitly identifying the important determinants of such movements in relative values.

Though constructed to address the special problems associated with some preservation-development decisions, the models can be used to address applied economic problems. Altering the patterns of demand and supply from those postulated for the environmental resources will broaden their applicability. These models do not claim to portray real-world processes completely; they are structured as guides to understanding the outcome of real-world processes.

SOME PRELIMINARY DEFINITIONS

The primary uses for the services of natural environments distinguished here are those of preservation and development. The consequences of these two alternative states must be understood. A natural environment is usually composed of complex ecosystems in which there has been minimal disturbance by man. Essentially unmodified, with natural elements predominant, such an area can provide unique recreational experiences quite different from those offered by commercial camping, hunting, and boating. Equally important, such areas are often the only preserve of the community's genetic stock, whose scientific value for agriculture, medicine, and many other fields may be very great (see Moir 1972). *Preservation* is that state which leaves the resource relatively untouched and in which natural forces dominate other influences.

Development, on the other hand, implies extensive modification of the environmental resources in question. It may be an extractive activity, such as mining or timber harvest, or it may transform the resource, as in damming a river for hydroelectric power. In either case, the results of the developmental activity are best described in terms of some extractive goods, such as electric power, minerals, or lumber.

Consequently, it will be convenient for our modeling efforts to distinguish the amenity services furnished by an environmental resource from the "direct services" of the resource. We shall assume that each environmental resource can provide a stream of homogeneous services that may be used either to produce amenity services or for the extraction of raw materials for production of alternative fabricated goods. These direct services from the resource will be affected by some activities and not by others. For example, clear-cutting a redwood forest for timber, one way of using the direct services of the resource, will make them unavailable to future generations. Hence this type of development represents an irreversible allocation of these direct services. Alternatively, the use of the forest's direct services by recreationists to produce for themselves amenity services (e.g., hiking, viewing the scenery) does not constitute an irrevocable allocation.

Although this distinction between direct services and amenity services may seem to be an artificial abstraction, it does provide a convenient means of integrating the problems associated with the allocation of environmental resources into the more traditional framework for economic analysis, allowing us to distinguish the substitutes for an environmental resource's direct services in the process of production from those goods or services that may substitute for amenity services in consumption. It is clear that as long as potential pollution externalities are not significant, the final consumer of electric power will be indifferent as to whether the power comes from a hydroelectric plant or a nuclear source. That is, the two alternative production modes provide outputs that are perfect substitutes in consumption. In contrast, there are usually few, if any, substitutes for the wilderness area's direct services in the production of low-density wilderness recreation and the associated amenity services.

TECHNOLOGICALLY INDUCED CHANGES IN RELATIVE PRICES AND BENEFIT-COST ANALYSIS

Over the past three decades, applied welfare economics has developed to guide public policy on intervention with market and nonmarket resource allocation in the United States. Since conventional economic analysis has established the perfectly competitive model as its norm or frame of reference for evaluating alternative resource allocations, market prices are taken to reflect the opportunity costs associated with any reallocation of resources; that is, under conditions of perfect competition in product and factor markets, prices reflect both the marginal consumer's willingness to pay for the product or service and the real costs of its provision.[2]

However, the use of market prices to reflect these resource opportunity costs assumes: (1) that there are no imperfections in the markets influencing determination of price; (2) that there is full employment of resources; and (3) that the individual project under evaluation does not affect the determination of product or factor prices. Although the last assumption may be valid for developed nations such as the United States, in the case of less-developed countries it can pose serious problems. Clearly both the extent of the market, or markets, involved and the magnitude of the project will jointly determine whether this assumption can be reasonably maintained.[3]

There has been substantial discussion of the extent to which prices are adequate measures of opportunity costs. The traditional approach, as presented by the Subcommittee on Evaluation Standards of the Inter-Agency Committee on Water Resources, assumes that relative prices remain constant over the life of the project, regardless of the

[2] Assuming that the given income distribution is acceptable. Krutilla (1961) has outlined the conditions under which public intervention is likely to achieve some overall improvement:

> Assuming that the gross benefit achieved (by public intervention to redirect the use of resources) exceeds the associated opportunity costs, if, in addition (a) opportunity costs are borne by beneficiaries in such wise as to retain the initial income distribution, (b) the initial income distribution is in some sense "best," and (c) the marginal social rates of transformation between any two commodities are everywhere equal to their corresponding rates of substitution except for the area(s) justifying the intervention in question, then welfare can be improved by such intervention.

[3] For a more complete discussion, see Eckstein (1961). Prices themselves are a function of the distribution of income. Any project which affects this distribution must face the issues raised by Kaldor, Hicks, Scitovsky, and Samuelson. There is no unambiguous way to determine whether or not there has been a net gain in societal welfare when some members benefit and others lose as a result of a particular action.

movement in the general price level. The subcommittee's report calls specifically for the exclusion of inflationary or deflationary movements in long-term average prices. Moreover, since the bond yield rate, a frequent choice for the discount rate, was thought to be markedly affected by inflationary or deflationary expectations in the bond market, the subcommittee recommended that a "deflated" discount rate be used so that these components would be approximately eliminated. The rationale for this policy stems from a recognition that inflation will increase the perceived monetary value of future benefits of a project without increasing their value in real terms.[4] Thus the objective is to attempt to measure the real benefits associated with the project under evaluation.

Within this framework it is important to distinguish between changes in the general price level, which the previous statements address, and relative price changes, which have received less attention. In the case of the former, the subcommittee's adjustments are roughly in keeping with measuring real benefits.[5] However, changes in relative prices do change the rate at which goods and services exchange and therefore can markedly affect our measures of real benefits. Krutilla and Cicchetti (1973) have argued that there are important reasons why a distinction must be drawn:

> The costs of extracting natural resource commodities and their market prices historically were shown to have remained either stable (for some) or actually declined (for others) relative to the price of goods and services in general. Accordingly, since these were the commodities which were being produced, in part, as outputs of the public resource development programs, there was in fact an authentic change in the price of outputs of such programs relative to the general price level. [pp. 60–61]

In those cases where public intervention involves a choice between preservation and development of unique natural environments, the movement of the relative values of the incompatible service flows is extremely important to the evaluation of the project. Analysis of such programs as the Alaska pipeline, the supersonic transport, or large-scale

[4] See the testimony of Henry P. Caufield, Jr., in *Economic Analysis of Public Investment Decisions: Interest Rate Policy and Discounting Analysis*, Hearings before the Subcommittee on Economy in Government of the Joint Economic Committee, 90 Cong., 2 sess. (Washington, D.C.: Government Printing Office, 1968), pp. 14–15; and *Proposed Practices for Economic Analysis of River Basin Projects*, report to Inter-Agency Committee on Water Resources, prepared by the Subcommittee on Evaluation Standards (1958), p. 20.

[5] However, given that the selection of a discount rate must be made to satisfy several objectives, the deflated rate may precipitate undesirable side effects. See Fisher and Krutilla (1974).

breeder-reactor development must reflect the price effects these expenditure decisions can be assumed to generate. The most important determinants of relative price changes are neither the size of the project and its redistributive effects on product or factor markets (which are likely to be marginal at a given time) nor the extent to which resources are fully employed, but demand and supply characteristics, including the inability of our productive technology to reproduce unique natural environments.[6] These resources, the result of an extremely long process of natural gestation, cannot be duplicated through man-made production processes. Since they are an essential input to the production of amenity services (i.e., recreational and other on-site activities), their existence is required for a variety of such service flows. Moreover, recent studies of the character of individual demand for recreation indicate that the structure of preferences does not allow other "produced recreation" to substitute readily for these amenity services (see Cicchetti 1972 and Stankey 1972). Accordingly, given increasing demand for amenity services and a continuing tendency of public resource allocation policy to favor conversion of our natural environments from their original to a developed state, we can anticipate progressive increases over time in the relative value of these amenity services.[7]

This monograph seeks to establish systematically those characteristics of both demand and supply which will influence such relative value appreciation. There are several important distinctions between the present model and the more popular neoclassical growth models. First, the models presented below describe the equilibrium points resulting from comparative static analysis. A dynamic equilibrium is traditionally defined for the neoclassical growth framework (e.g., Harrod neutral growth path). Second, the nature of the demand structure in the neoclassical framework is crude at best and assumes that the quantity demanded will increase as a constant fraction of real income;[8] our model allows for complex demand structures in which both income and substitution effects are important. Finally, because the neoclassical models are largely concerned with the factor markets, the equilibrium is defined in terms of the capital-labor ratio. The present model is concerned with the product markets, particularly with the relative prices that a perfectly competitive market would provide; consideration is

[6] Irreversibilities do, however, have significant intertemporal implications.

[7] This tendency seems to run in direct opposition to the reasoning which presumably underlies the public subsidies associated with natural resource extractions, such as the depletion allowance for petroleum.

[8] There are models which allow for differences in the savings propensities as between workers and rentiers. See Burmeister and Dobell (1970) for a summary.

therefore given to factor market effects *only* to the extent that they might influence the movement in relative values (i.e., prices in perfectly competitive markets).

In spite of the development of a substantial body of valuable literature in growth theory, there seems to be no simple scheme within any of the traditional models to account for the determinants of relative price movements in reponse to technological change and an important intertemporal externality—when a resource allocation at one time permanently affects the possibility of alternative allocations in the future. An important start has been made by Baumol (1967), who focused on the outlook for the arts and the provision of public services in an economy with technologically more progressive sectors. His models were consequently greatly simplified and illustrative. Baumol eliminated the demand side of a general equilibrium model by postulating a structure where relative prices were determined solely by supply. We shall adapt Baumol's basic approach to the environmental problem and develop a complete specification of the demand side of the model.

The model developed in the following chapters is based on several simplifying assumptions. First, I have assumed that the community's supply options can be described by a production possibility curve with a traditional neoclassical contour. This assumption necessarily implies that factor markets are not sufficiently imperfect to distort the smooth concavity of the frontier. At first this requirement may seem overly restrictive, since one of the goods in question is the amenity services of natural environments. However, it is reasonable when we recognize that these amenity services are produced by combining the "direct" services of natural environments with other factors, such as an individual's time, and that there is little scope for substitution for the direct services in this production process. The reasoning underlying this statement is straightforward.

The extent to which factor market imperfections, such as those which might result from the character of the direct services of natural environments (e.g., nonmarketability, difficulty of exclusion, etc.), can affect the convexity of the production possibility set is contingent upon the technical parameters of the processes providing the goods and services mapped in that transformation curve. That is, the nature of productive technology will determine whether these factor market imperfections severely affect the loci of community options. In terms of the parameters used to describe the production processes involved, small elasticities of substitution between the factor in question and other inputs will tend to minimize the impact of the imperfection. Thus, for our specific case, the social production set will be likely to satisfy the convexity requirements for an equilibrium.

The second major assumption rests on the characteristics of community demand and therefore on the definition of convenient, yet representative, vehicles for describing it. Consequently we can assume, without loss of generality for our specifically defined objectives, that a utility function with a preassigned pattern of demand parameters adequately describes the community's preferences. We thus avoid the issue of the relationship of the individual to the whole and the aggregation of preferences.

Opportunity costs reflect both the trade-off implied by the nature of the supply conditions (i.e., the marginal rate of transformation) and that from the demand side (i.e., the marginal rate of substitution). Benefit-cost analysis has traditionally assumed that we are (with the exceptions noted) evaluating the opportunity costs associated with resource reallocations at a point of equality of these trade-offs in production and consumption. The models developed herein examine how the equilibrium rate of substitution, which is our appropriate measure of opportunity costs, changes with exogenous changes in technology and with intertemporal externalities.

Technical innovation may not have the same effect on the different uses to which an environmental resource may be put, and there is reason to suspect that differential rates of technological change will be experienced by different sectors. Thus there is an asymmetry in the implications of technological change for the value of environmental resources when their services are used for mutually exclusive purposes.

Three considerations are important to the magnitude of technologically induced changes in relative prices. First, for any two goods, such as a fabricated good and the amenity service, the income elasticity of demand for the latter relative to the former will affect relative price movement. If, for example, there is no appreciable difference in how demand for these goods responds to income changes, then increases in the equilibrium consumption of amenity services will not contribute greatly to relative price changes. Moreover, in this case, the consequences of irreversibilities will be negligible. Second, the respective availability of substitutes for each of the commodities in question is also important. Consider, for example, the case in which there is at least one reasonably good substitute (measured by the partial elasticity of substitution in consumption) for the amenity services. This condition will, *ceteris paribus*, tend to offset the consequences of the differential in the incidence of technical change (which I have taken as a given).

Finally, we must evaluate the ability of the economy to transform its resources, both at a point in time and through time, in order to produce alternative combinations of commodities. That is, the social production set describes at any particular time the alternative combinations of goods and services that the community may choose. It is based upon a fixed resource base and given technology. The shape of the contour reflects how easily the resources may be converted from the production of one good to another. Consequently, the curvature reflects the importance of the changes in relative costs, as a result of the law of diminishing returns in production.

Irreversibilities are one type of intertemporal production externality and as such do not affect the loci of production possibilities at any given time. Rather, the production locus at any one time is a function of the commodity mix chosen in past periods. For example, it was deemed desirable, for a variety of reasons, to construct a dam to flood Glen Canyon. Today's recreationists cannot explore the unmodified canyon or enjoy white-water recreation as it existed on the Colorado River before the dam was built. We cannot reverse this choice. Future generations must accept the more constrained set of options that the present will leave to them. While they will have access to more power as a result of the hydro-facility, they will have been pre-empted from enjoying the canyon's amenity services. In this way, the choices of commodity mixes over time will affect the rate at which relative prices change.

The models developed in the following chapters allow for some more practical analysis as well. Fisher, Krutilla, and Cicchetti (1972) have developed a model capable of measuring what the initial year's preservation benefits *need to be* to equal the present value of net benefits from a developmental action. Their model assumes that the demand for the amenity services from preserved natural endowments will increase progressively over time. The growth in benefits is then linked to the forces influencing growth in demand (in their case, shifting of a linear demand curve). One source of growth is a postulated increase in the relative price at every quantity level as a result of income growth due to technological change. If we are willing to assume that the relative prices of all quantities move as do equilibrium exchange rates, then for given demand parameter configurations (i.e., income elasticities and cross-price elasticities for all goods), it is possible to associate this rate of relative price change with the growth in technical change. Of course, as I noted at the outset, the specification of these models has not been completely general, and therefore we must accept some approximation, due to the forms of the specific utility and transformation functions.

However, our models allow the definition of a set of feasible values for this rate of relative price increase, so that the benefit estimation described by Fisher, Krutilla, and Cicchetti can be directly linked to the character of demand for each case of interest.

THE PLAN OF THE STUDY

Economists have devoted a great deal of attention to modeling and measuring technical change. Although we are interested in the effects of technological change rather than in modeling its movement, the primary contributions and shortcomings of such studies must be clearly understood. The second chapter, a review of the literature, defines a productive technology and its relationship to the production function, develops the rudiments involved in the modeling of technological change, and explains the deficiencies of most of these models to the problems under discussion.

The models developed in chapters 3–6 are simple general equilibrium models. In many respects they handle the most difficult questions generally associated with such models by simplifying assumptions.

Chapter 3 reviews the assumptions underlying the construction of production possibility frontiers, which we use to summarize the supply side. The "special" character of the "direct services" of environmental resources is considered. That is, because factor markets cannot correctly price the services of common property resources, they may not always be allocated efficiently. This imperfection can affect the traditional shape of the transformation curve, thus necessitating consideration of existence and stability of the equilibrium positions for the model.

After a brief review of the characteristics of demand for amenity services, chapter 4 outlines the mode used to represent community demand. The community indifference curve, a convenient vehicle for depicting aggregate preferences, is used; the link between it and the structure of demand is illustrated using the Hicks-Allen equations. Clearly the choice of specification for the community indifference map will affect the pattern of price, income, and cross elasticities of demand. To illustrate the importance of this choice, the demand structures implied by several possible specifications for the community indifference curves are reviewed. Overall, the flexibility of the Mukerji (1963) constant ratio of elasticities of substitution makes it the most desirable alternative.

The tools developed to describe demand and supply are then used in chapter 5 to examine the character of the time path of relative prices under conditions of autonomous technical change favoring fabricated commodities. These time paths describe the prices of amenity services relative to those of fabricated goods at the comparative static equilibrium positions. Both two- and three-good models are developed and alternative utility specifications examined, so that one can gauge the impact of the structure of demand and of the generality of the model upon the implied movement in relative prices.

Chapter 6 extends these models, allowing for an important type of intertemporal externality, irreversibilities. By assuming that an allocation of the direct services of an environmental resource is irreversible, this chapter demonstrates the effect of increased absolute scarcity on relative price behavior. The last chapter summarizes the important considerations in our models and the implications of their results.

REFERENCES

Barnett, H. J., and C. Morse. 1963. *Scarcity and Growth*. Baltimore: The Johns Hopkins Press.

Baumol, W. J. 1967. "Macroeconomics of Unbalanced Growth: The Anatomy of Urban Crisis." *American Economic Review* 57: 415–426.

Burmeister, E., and A. R. Dobell. 1970. *Mathematical Theories of Economic Growth*. New York: Macmillan Co.

Cicchetti, C. J. 1972. "A Multivariate Statistical Analysis of Wilderness Users in the United States." In *Natural Environments: Studies in Theoretical and Applied Analysis*, ed. J. V. Krutilla. Baltimore: The Johns Hopkins University Press for Resources for the Future, Inc.

Clawson, M., and J. L. Knetsch. 1966. *Economics of Outdoor Recreation*. Baltimore: The Johns Hopkins Press.

Eckstein, O. 1961. "A Survey of the Theory of Public Expenditure Criteria." *Public Finances: Needs, Sources, and Utilization*. Princeton, N.J.: Princeton University Press.

Fisher, A. C., and J. V. Krutilla. 1972. "Determination of Optimal Capacity of Resource-Based Recreational Facilities." *Natural Resources Journal* 12: 417–444.

———. 1974. "Resource Conservation, Environmental Preservation, and the Rate of Discount." *Quarterly Journal of Economics*. Forthcoming.

Fisher, A. C., J. V. Krutilla, and C. J. Cicchetti. 1972. "The Economics of Environmental Preservation: A Theoretical and Empirical Analysis." *American Economic Review* 62: 605–619.

Harberger, A. C. 1971 "Three Basic Postulates for Applied Welfare Economics: An Interpretive Essay." *Journal of Economic Literature* 9: 785–798.

Krutilla, J. V. 1961. "Welfare Aspects of Benefit Cost Analyses." *Journal of Political Economy* 69: 226–235.

———. 1967. "Conservation Reconsidered." *American Economic Review* 57: 777–786.

Krutilla, J. V., and C. J. Cicchetti. 1973. "Benefit-Cost Analysis and Technologically Induced Relative Price Changes: The Case of Environmental Irreversibilities." In *Benefit Cost Analyses of Federal Programs*. Joint Economic Committee. Washington, D.C.: Government Printing Office.

Moir, W. H. 1972. "Natural Areas." *Science* (August 4, 1972).

Mukerji, V. 1963. "A Generalized SMAC Function with Constant Ratios of Elasticity of Substitution." *Review of Economic Studies* 30: 233–236.

Potter, N., and F. T. Christy, Jr. 1962. *Trends in Natural Resource Commodities: Statistics of Prices, Output, Consumption, Foreign Trade, and Employment in the United States, 1870–1957*. Baltimore: The Johns Hopkins Press for Resources for the Future, Inc.

Stankey, G. H. 1972. "A Strategy for the Definition and Management of Wilderness Quality." In *Natural Environments: Studies in Theoretical and Applied Analysis*, ed. J. V. Krutilla. Baltimore: The Johns Hopkins University Press for Resources for the Future, Inc.

chapter two Modeling the Effects of Technical Change

The methods we shall be using should be put into perspective against the conventional approaches for analyzing the effects of technical change. Definitions of the production function, factor inputs, and technical change itself are reviewed below. A brief survey of traditional approaches suggests that they will not be particularly useful for our purposes. They tend to focus on the factor input effects of changes in the productive technology, whereas our concern is with the implications of these changes differentially across product markets.[1] The models used here are indebted to Baumol's (1967) unbalanced growth model, which is reviewed below.

SOME DEFINITIONAL ISSUES

The term *productive technology* describes the technical means whereby materials and services (from labor, man-made capital, or the environ-

[1] Detailed reviews of the literature are available in Nadiri (1970), Kennedy and Thirlwall (1972), and Smith (1973).

ment) may be combined to produce some commodity or service (Shephard 1970, p. 13). The goods and services used in the process of this production are called the factor inputs. This definition does not imply a simple functional relationship linking the factor inputs to the outputs of the process. More often than not, the technology is quite complex, and simplistic descriptions fail to capture all the attributes of the actual system (see Russell 1973). Nonetheless, economists have approximated complex technological relationships with relatively simple functional forms linking broadly defined factor inputs (such as labor and capital) and the output.[2]

Brown (1968) notes that a production function "enters economic analysis as a datum—given by technology or extra-economic considerations" (p. 9). We assume that the entrepreneur knows how to produce the maximum output with a given combination of inputs. By assumption, the production function specifies the maximum output for each combination of inputs.

To use the production function abstraction properly, one must recognize that: (1) the function is a convenient formulation for describing a more complex underlying production process, and (2) the technical efficiency and extraeconomic characteristics of the function are assumed. An observationally equivalent explanation of behavior need not operate with the same postulates.[3]

Conventional economic analysis displays the production function graphically in a variety of modes. One of the most popular is the constant output, or isoquant, representation of the function. Any one of these curves describes the loci of factor input combinations that will yield the same output. This representation is important to micro-economic

[2] Unfortunately, in their approximation economists until recently (see Ayres and Kneese 1969) have neglected the nonmarketable outputs resulting from their production processes. These industrial residues were omitted because costless disposal mediums were available. Today the ambient environment is threatened and society is more responsive to schemes requiring firms to internalize the costs of disposal of these residuals.

[3] These points will become increasingly important when induced technical change is discussed. As Salter (1969) notes:

> By selecting methods to be developed in detail, in the designing process itself, and in the choice of equipment to be designed and marketed, engineers and machine-makers anticipate the needs of businessmen. The analytical problem is not so much this division of labor in the choice of techniques; if all decisions are made according to the same principles the net result will be much the same as if they were centralized. Rather the problem is how we should think of a range of alternative techniques of production. This is a problem that can be expressed in terms of the production function, the traditional means of describing alternative methods of production in terms of the required inputs of factors of production. The above argument implies that there are two ranges of alternative techniques and the production function could refer to either. [p. 14]

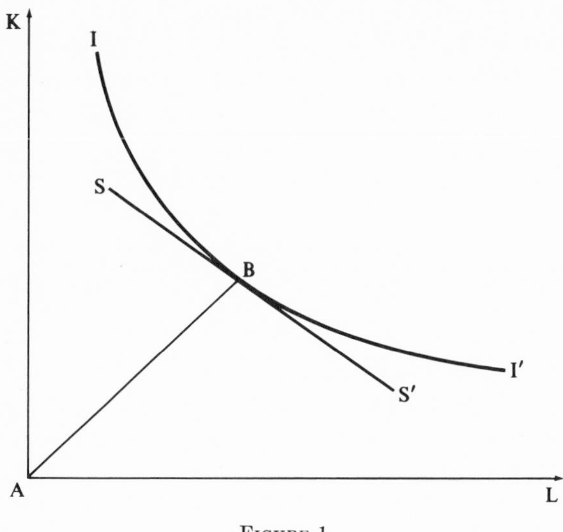

FIGURE 1.

analysis because it allows one to define those characteristics of the abstract technology which are important to the input resources' utilization. Brown (1968, pp. 13–20) has outlined four characteristics of an abstract technology and has discussed how each affects the defined production function. They include efficiency, economies of scale, capital intensity, and the ease with which capital is substituted for labor (i.e., elasticity of substitution). The first of these measures the factor mix required to produce a given output. The scale factor refers only to technical characteristics of the production process, not to market advantages as a consequence of size.[4] Since the usual neoclassical model is a two-factor model, capital intensity is synonymous with the input mix. The last attribute of an abstract technology, the elasticity of substitution, is a measure independent of units and can be interpreted as an indication of the curvature of a given isoquant. The larger the elasticity, the flatter the isoquant. A Leontief technology would assume a zero elasticity of substitution; one with perfectly substitutable factor inputs would have an infinitely large value for the elasticity.

Figure 1 illustrates the definition in terms of the rates of change of two slopes. The elasticity of substitution measures the rate of change of the slope ray (AB) from the origin to a point on the isoquant (II'), say point B, relative to the rate of change in the slope of a cord (SS') tangent

[4] Russell (1973) notes some of these issues with regard to an oil refinery.

to the isoquant at the same point B.[5] In economic terms, it measures the percentage change in the capital-labor ratio relative to the percentage change in the marginal rate of substitution of capital for labor along a given isoquant. If we define the elasticity, σ, in equation (1), then it is possible to derive the relationship between the capital-labor ratio and the marginal rate of substitution as σ changes.

$$\sigma = \frac{dy}{dx} \cdot \frac{x}{y} \qquad (1)$$

where: K = capital
L = labor
$y = K/L$
$x = MP_L/MP_K$
MP_K = marginal product of K
MP_L = marginal product of L

Suppose at B σ is a constant, σ^*, then we can rewrite (1) as (2):

$$\frac{dy}{y} = \sigma^* \frac{dx}{x} \qquad (2)$$

Integrating both sides, we have:[6]

$$\ln y = \sigma^* \ln x + c \qquad (3a)$$

where: c = constant of integration

Substituting for y and x, we have:

$$\ln K/L = \sigma^* \ln \frac{MP_L}{MP_K} + c \qquad (3b)$$

This equation allows us to illustrate two of the four characteristics of an abstract technology. Changes in c alter the capital intensity for given σ^* and given MP_L/MP_K (i.e., marginal rate of substitution of K for L). In addition, we can allow σ^* to change, altering the relationship between K/L and MRS_{KL} at all points.

With these definitions it is possible to discuss the conventional modeling of technological change. Technological change implies that the productive technology underlying the economists' production function has altered, that more goods and/or services (outputs) can be obtained with little or no variation in those resources used to produce them. In terms of the isoquant description, the mapping of constant product curves has moved closer to the origin. Whether that movement has been parallel for

[5] See Smith (1969) for derivation of the *CES* production function following the same reasoning.

[6] Brown (1968) provides a similar argument and a slightly different derivation. See his appendix A.

all capital-labor mixes will be discussed below. The principal point is that the new technology's processes for producing a given output level require less of at least one factor input, with others unchanged, than did the old process.

Our definition relates only to our production function construction. It is a translation into abstract terms of the changes that may have occurred in the production technology underlying production functions.[7] Moreover, our only means for describing the change is in terms of the four characteristics of a technology we have defined. Such description does not explain the causes of change. Nor does it necessarily mean the changes were made by entrepreneurs after viewing the abstract economic functions. Rather it implies that their motivation and actions are consistent with the assumptions underlying these functions. As a consequence, observationally equivalent explanations of entrepreneurs' behavior may require alternative models for accounting for observed changes in the production technology.[8]

CONVENTIONAL METHODS FOR MODELING TECHNICAL CHANGE

The abstractions (i.e., production functions) central to economists' modeling of complex productive technologies have tended to focus on how change in these technologies affects the goods and services defined as inputs to the respective processes. In order to determine the implications of technical change for the outputs of these processes, it is necessary to enumerate its effects on each factor in each production process and to work through the specified production functions. Since regressive technical changes (i.e., those which reduce output with given inputs) have been assumed absent with traditional models, changes in the technology provide, by definition, more of the output involved. The interesting problems were thought to be those associated with input effects.

Neutral and non-neutral technical changes have thus been defined in terms of the influence of the technical change on the respective factor inputs. In the most general sense, a parallel shift in the isoquant mapping, such as that from I to I' in figure 2, might be considered to depict a neutral change. It decreases the quantity of inputs (assuming the output associated with I equals that associated with I') required to produce

[7] Brown (1968) has this same point in mind by defining an *abstract* technology.

[8] This point is particularly relevant to the empirical testing of the models in question.

a given output, but does not change the marginal rate of substitution between them at each input mix. Of Brown's four characteristics of an abstract technology, only changes in efficiency and economies of scale represent neutral changes.

A non-neutral change increases the amount of output but affects each input's usage differentially. Such a change might be represented by a non-parallel shift in the map of isoquants, as in figure 3. Unfortunately, some confusion enters the literature relating to the neutrality of technical change. Micro and macro concepts are not clearly distinguished, and the definition of neutrality becomes obscured by other considerations. Three measures of neutrality have developed in the analysis of the aggregate

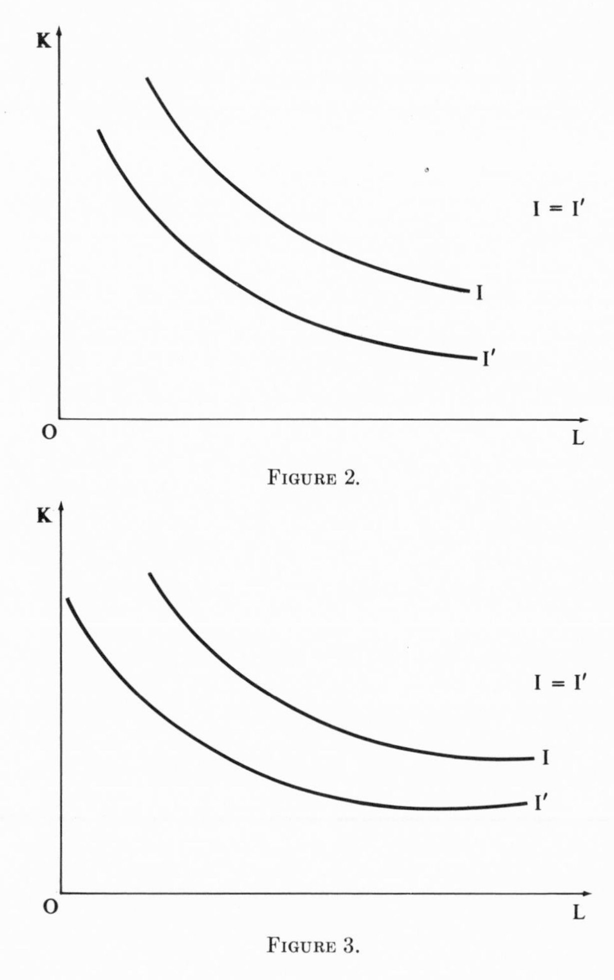

FIGURE 2.

FIGURE 3.

effects of technical change on factors. Their usefulness at the micro level is therefore questionable. Nonetheless, use of the terms persists. Hicksian neutral changes have been defined as those which do not affect the relative shares in a two-factor model for a given input mix (i.e., capital-labor ratio). This specification for neutrality was constructed to gauge the effects of technical change on the factors' respective shares of the output, not to describe the character of the change in the abstract technology. Similarly, Harrod neutral change was defined for convenient analysis of equilibrium growth paths, in neoclassical terms, of the constancy of relative shares for a given capital-output ratio. Solow neutrality, defined for completeness, calls for invariance of relative shares for a given labor-output ratio.[9]

Although considerable attention has been given to the relationship between these measures of neutrality and their respective effects on (1) the factor's share of the total output (in aggregate terms) and (2) the stability of neoclassical growth paths, little has been done to assess their adequacy in describing the changes in the underlying productive technology. Moreover, the bulk of theoretical research in this area has focused upon macro behavior with an assumed aggregate production function. Thus, though this literature has enlarged our understanding of the problems associated with aggregation and the definition of aggregated stock variables, it has not appreciably advanced the modeling of technical change for micro-economic problems.

The models to be developed here will not explain the process of technical change either, but will instead analyze its effects upon the prices of final goods and services. Thus the neutrality of primary interest in our models deals with the effect of technical change on outputs. Our concern is with which sector or sectors are most benefited by technology, and which least. Scenarios can be developed in terms of factor-related definitions of the changes, and the results for the outputs of interest can be derived. However, the combinations of requisite assumptions about the following expand rapidly: (1) the character of the specific production functions (i.e., elasticities of factor substitution, returns to scale, factor intensities), (2) the allocation mechanism for factor inputs, and (3) the basis for the definition of technical change. Since we do not have the information necessary to make these assumptions in other than an arbitrary manner and since we will be concerned ultimately with those simplifications which are important to the final outcome, a different strategy may be more appropriate.

[9] These three measures are best defined in Nadiri (1970).

The models proposed here will postulate a pattern of technical change in terms of its effects on final goods and services, without specifying what this pattern may imply for the factors used in obtaining the final outputs. Consequently, neutral and non-neutral changes can be defined in terms of outputs rather than inputs. Since the analysis will focus on comparative static equilibrium points, the adjustment process necessary to assure equilibrium, the stability of the path, the mix of factor inputs utilized, and other traditional considerations of neoclassical growth models are not discussed. We will look instead at the behavior of relative prices and how that behavior is influenced by measures of community demand and productive ability. In simplifying our model, we must be aware of the inherent limitations with this approach as well. Moreover, previous statements notwithstanding, the character of the inputs themselves will affect both the conventional approaches to the modeling of technical change and the present strategy. That is, if we must allow one of the factor inputs to be the services of a common property resource, then the implications need to be understood. Market characteristics have no effect upon the representation of production processes in production functions.[10] This statement should not be confused with one which calls for invariance in the behavior of the rational entrepreneur in the presence of such imperfections. There should be a clear distinction between the production function, which has been assumed to be technical data, and the choices which are made as a result of it, market information, and a set of relevant constraints, including the market imperfections associated with common property resources.

Thus, omitting the services of common property resources from the production functions in conventional models of technical change presupposes that their uses do not change. That is, it assumes that entrepreneurs do not consider, in either their short-run or long-run decisions, the possibility of substitution between factor inputs other than capital and labor. This point is especially important to both the modeling and the measurement of technical change. Schemes that classify types of innovations into neutral and non-neutral, using alternative definitions (i.e., Hicks, Harrod, or Solow), presuppose two factor inputs. Solow (1967), discussing the value of the factor augmentation representation of technical change, notes that:

the extra flexibility [of this approach] is especially valuable if one has to account for more than two factors of production. Of course, the whole analysis becomes

[10] Shephard (1970). He notes specifically that "free goods or services are not excluded as factors of production, since market prices have no bearing upon the technical roles of these inputs" (p. 13).

more complicated; one needs an index of the rate of technological progress plus two indexes of bias plus three elasticities of substitution, and there is even some choice about how to define the elasticity of substitution. Besides, if the context is economic growth, the only possible extension of the underlying idea of Harrod-neutrality turns out to be extremely limiting. One must suppose that there is only one primary factor, all the rest being themselves produced. If one requires that the average products of the produced factors all be constant and independent of the level of technology when all their marginal products are constant, then technological progress must augment only the single primary factor. [pp. 31–32]

One will surely grant, as Professor Solow suggests, that the analysis is much less "tidy." Moreover, the answers available with two-factor models may not be available with three factors, or an even greater number of inputs. The question is: what do the answers mean?

If we inquire into their meaning in the presence of a nonpriced factor input, our answer hinges upon the degree to which decisions upon factor employment are joint. That is, can the services of the third factor be considered substitutes for those of the other two?[11] If so, and the third factor is nonpriced, the measures of technical change in terms of factor shares constitute indexes in a partial sense only. Moreover, the interpretation of neutral and non-neutral changes by these definitions is not clear. Extreme non-neutralities may be present when a particular index indicates neutral change, or the dominant type of non-neutrality may not be revealed.

Over the past decade the increased demands placed upon the assimilative capacity of the environment reflect to some extent these neglected substitutions. Concomitant with this increase in industrial demands for environmental services has been an increasingly rapid growth in the demand for the amenity services of environmental resources. Our models need to be adapted to accommodate these resources.

The induced technical innovations models, particularly those on a micro-economic scale (see Kamien and Schwartz 1968, 1969), apparently can be most easily transformed to accommodate environmental resources. These models postulate that the path of innovations is related to the direction of movement in, and the levels of, relative factor prices.

[11] One question which may be occurring now is how one operationalizes these simple two-factor models so that estimation is possible. Economists do recognize that a variety of primary and intermediate goods enter into the production of any specific commodity. This question is resolved by measuring only the value added to that of primary and intermediate goods. Consequently empirically estimated production functions explain *only* the contribution of labor and capital to this addition to market value.

The nonpriced character of common property resources prevents our accounting for their contribution in a similar manner.

Accordingly, one suspects with Ruttan (1971), and can demonstrate in simple cases (see Smith 1972), that the "effect of continued undervaluation of environmental services has been to induce a pattern of technical change which is biased in the direction of excess residual production and away from increased efficiency in the supply of resource amenities" (p. 712). Ruttan indicates that this pattern is clear in agriculture. Innovation has been biased toward developing land substitutes (as a result of the high relative price for land) in the form of plant nutrients, crop varieties, and management systems, which, taken together, lead to a degradation of the environment.

Few models of either an autonomous or induced character can readily accommodate common property resources. Moreover, though there is increasing recognition of the importance of the problems of incorporating these resources into the models (see Ruttan 1971 and Nadiri 1970), our brief summary indicates that substantially different approaches to the modeling of technical innovation will be necessary before common property resources can be fully integrated into the explanatory mechanisms. This limitation, however, poses no problem to the objectives of the present research. It is not necessary to explain the "why" of a particular pattern of technological change. Thus the unique characteristics of common property resources are important to the analysis only in the extent to which they affect the tools of our analysis; these implications will be developed in chapter 3.

BAUMOL'S LEGACY: MACRO-ECONOMICS OF UNBALANCED GROWTH

As I noted in the first chapter, the approach utilized in the present analysis owes a great deal to the unbalanced growth model developed by W. J. Baumol (1967). His model will be briefly reviewed here as a starting point for the analysis in subsequent chapters.

Baumol's model is structured on the basis of a two-product, one-factor model with production functions as given in equations (4) and (5).

$$Y_{1t} = aL_{1t} \tag{4}$$

$$Y_{2t} = be^{rt}L_{2t} \tag{5}$$

Output (Y) is assumed to be a constant multiple of the labor resources (L) used in each sector at each moment in time. Moreover, the productivity of labor in the production of Y_2 is assumed to be increasing at a constant compounded rate, r. Sector two is therefore the technologically progressive sector.

Assuming wages are equal in both sectors and increasing at the rate of productivity increase in the second sector, Baumol derives four propositions. They may be summarized as follows: (1) the unit cost of Y_1 will rise without limit, while that of Y_2 will remain constant; (2) there is a tendency for Y_1 to decline, even to disappear, if demand is not highly inelastic; (3) if the Y_1/Y_2 ratio is to be held constant, increasing fractions of the total labor force must be diverted to the production of Y_1; and (4) any attempt to balance growth with unbalanced productivity must reduce the rate of growth relative to the rate of growth of the labor force.

One of the most valuable aspects of Baumol's paper is his thoughtful discussion of those areas in which the model appears adequately to describe the existing conditions. However, the model itself does serve to illustrate one simple methodology for determining the effects of technical change in the product markets. Equally important, it provides a bench mark for measuring the importance of both demand and supply effects on the behavior of the relative prices and commodity mix ratios at the equilibrium point. If we assume a fixed supply of labor, L_t, which is fully employed at any moment in time, then the production possibility curve may be derived as follows. Solving (4) and (5) for L_{1t} and L_{2t}, we have:

$$L_{1t} = \frac{Y_{1t}}{a} \qquad (6)$$

$$L_{2t} = \frac{Y_{2t}}{be^{rt}} \qquad (7)$$

If the total resources are fully employed, then (8) must hold.

$$L_t = L_{1t} + L_{2t} \qquad (8)$$

Substituting from (6) and (7) into (8) yields the production possibility curve in (9).

$$\frac{1}{a} Y_{1t} + \frac{1}{be^{rt}} Y_{2t} = L_t \qquad (9)$$

It should be noted that Baumol's model provides a production possibility curve with a constant marginal rate of transformation, thereby assuming no diminishing returns in the production of Y_1 and Y_2, as is apparent from their respective production functions in (4) and (5). Figure 4 illustrates the nature of the frontier. Supply completely determines relative prices in this model. Moreover, since Baumol's model provides no formal specification of demand, it is not possible to examine the off-

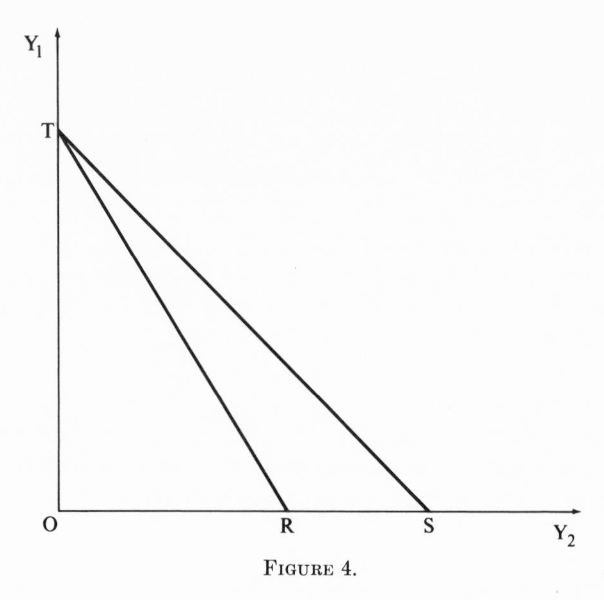

FIGURE 4.

setting effects of preferences versus technological change in the determination of the equilibrium commodity mix.

Technical change in sector two serves to pivot the production possibility curve about the point aL_t (point T in figure 4) in the Y_1 coordinate. The movement from TR to TS in figure 4 illustrates the effect. The relative prices of Y_{1t} and Y_{2t} change with the slope of the frontier, and can be described in equation (10).

$$\frac{P_{1t}}{P_{2t}} = \frac{b}{a} e^{rt} \tag{10}$$

Determination of the time path for the commodity mix (i.e., Y_{1t}/Y_{2t}) requires the formal specification of the community demand. As noted, Baumol does not extend his model to include completely specified demand, but discusses instead the implications of alternative demand patterns for the model, which has led to some confusion in the comments on his paper. Worcester (1968), for example, remarks:

> . . . his analysis suggests the less likely of at least two logical conclusions, namely that the production in the less progressive sector will tend to diminish. This may reflect Baumol's concentration on the price and income elasticities of goods in the unprogressive sector and to his failure to introduce a specific community indifference curve. . . . under the technological conditions postulated any of a large family of community indifference curves will eventually place the whole labor force into the unprogressive sector, and it is by no means clear that this family is a less likely eventuality. [pp. 887–888]

The pattern of community demand elasticities is not independent of the specification of a community indifference mapping whether for the individual or the community. This mapping determines a specific structure of price, cross, and income elasticities. (Chapter 4 takes up this point in some detail.)

Our models will differ from Baumol's structure in two major respects. First, the production possibility curves allow for diminishing returns in the production of each final good, so that both demand and supply factors influence the path of equilibrium relative prices. Second, specific demand structures will be specified so that the influence of tastes may be gauged against that of technological considerations.

The problems we shall address include (1) the differential effects of technology on fabricated goods relative to the amenity services of natural endowments, and (2) an intertemporal externality resulting from the use of natural endowments for developmental purposes.

SUMMARY

A brief review of the definitional issues underlying conventional methods for modeling productive technology has allowed us to compare traditional ways of looking at the implications of technical change with that of this monograph. Several points have emerged.

(1) A production function is an approximation assumed to capture the essential characteristics of a more complex underlying productive technology.

(2) Brown (1968) has defined four characteristics of an abstract technology: efficiency, economies of scale, capital intensity, and the elasticity of substitution of capital for labor.

(3) Technical change means simply that the productive technology underlying the economist's production function has changed. Most models of technical change have sought to describe these changes rather than explain them.

(4) In a general sense, neutral technical change implies a parallel shift in the isoquant map toward the origin, and this, in fact, is a legitimate representation. However, in the definition of non-neutral change, three measures of neutrality have been devised: Hicks, Harrod, and Solow.

(5) Induced technical change models seek to make the direction of innovation an endogenously determined variable. Most models relate this direction to the value expected or changes in the value of relative

factor prices. They do not, however, explain the allocation of resources to, or the production process of, inventive activity.

(6) The importance of common property resources to models with autonomous technical change hinges upon the degree to which factor hiring decisions are joint. Moreover, the interpretation of traditional measures of the non-neutrality of technical change is not clear. These measures may in fact be seriously misleading.

(7) Baumol's unbalanced growth model provides the starting point for our analysis. The present models have a completely specified demand structure and allow for diminishing returns in production. His model did not account for either of these factors.

REFERENCES

Ayres, R. U., and A. V. Kneese. 1969. "Production, Consumption, and Externalities." *American Economic Review* 59(3): 282–297.

Baumol, W. J. 1967. "Macroeconomics of Unbalanced Growth: The Anatomy of Urban Crisis." *American Economic Review* 57: 415–426.

Brown, M. 1968. *On the Theory and Measurement of Technological Change*. Cambridge: Cambridge University Press.

Burmeister, E., and A. R. Dobell. 1970. *Mathematical Theories of Economic Growth*. New York: Macmillan Co.

Fellner, W. 1971. "Empirical Support for Induced Innovations." *Quarterly Journal of Economics* 85: 580–604.

———. 1967. "Technological Progress and Recent Growth Theories." *American Economic Review* 57(5): 1073–98.

Ferguson, C. E. 1969. *The Neoclassical Theory of Production and Distribution*. Cambrige: Cambridge University Press.

Fisher, F. M. 1969. "The Existence of Aggregate Production Functions." *Econometrica* 37: 533–577.

Gruen, F. H. 1961. "Agriculture and Technical Change." *Journal of Farm Economics* 43: 838–858.

Hayami, Y., and V. W. Ruttan. 1970. "Factor Prices and Technical Change in Agricultural Development: The United States and Japan, 1880–1960." *Journal of Political Economy* 78: 1125–35.

Hicks, J. R. 1965. *Value and Capital*. 2nd edition. Oxford: Oxford University Press.

———. 1968. *The Theory of Wages*. 2nd edition. New York: St. Martin's Press.

Kamien, M. I., and N. L. Schwartz. 1968. "Optimal 'Induced' Technical Change." *Econometrica* 36(1): 1–17.

———. 1969. "Induced Factor Augmenting Technical Progress from a Microeconomic Viewpoint." *Econometrica* 37(4): 668–684.

Kennedy, C., and A. P. Thirlwall. 1972. "Surveys in Applied Economics: Technical Progress." *Economic Journal* 82: 11–68.

Nadiri, M. I. 1970. "Some Approaches to the Theory and Measurement of Total Factor Productivity." *Journal of Economic Literature* 8: 1137–77.

Nordhaus, W. D. 1969. *Invention, Growth, and Welfare: A Theoretical Treatment of Technological Change*. Cambridge, Mass.: The M.I.T. Press.

Russell, C. S. 1973. *Residuals Management in Industry: A Case Study of Petroleum Refining*. Baltimore: The Johns Hopkins University Press for Resources for the Future, Inc.

Ruttan, V. W. 1971. "Technology and the Environment." *American Journal of Agricultural Economics* 53(5): 707–717.

Salter, W. E. G. 1969. *Productivity and Technical Change*. 2nd edition. Cambridge University Press.

Schmookler, J. 1960. *Invention and Economic Growth*. Cambridge, Mass.: Harvard University Press.

Shephard, R. W. 1970. *Theory of Cost and Production Functions*. Princeton, N.J.: Princeton University Press.

Smith, V. K. 1969. "The CES Production Function: A Derivation." *The American Economist* 13(1): 72–76.

———. 1972. "The Implication of Common Property Resources for Technical Change." *European Economic Review* 3: 469–479.

———. 1973. "A Review of Models of Technological Change with Reference to the Role of Environmental Resources." *Socio-Economic Planning Sciences* 7: 489–509.

Solow, R. M. 1967. "Some Recent Developments in the Theory of Production." In *The Theory of Empirical Analysis of Production*, ed. M. Brown. Studies in Income and Wealth, NBER.

Worcester, D. A., Jr. 1968. "Macroeconomics of Unbalanced Growth: A Comment." *American Economic Review* 58: 886–895.

chapter three The Supply Side: Technically Feasible Production Possibilities

In the absence of product market imperfections, the technical options available to a society can be conveniently summarized by a production transformation curve. In this chapter, the production transformation curve is defined and derived graphically as well as algebraically. Moreover, the effect of environmental resources upon the shape of the contour is discussed. Neutral and non-neutral, autonomous technical innovations in terms of products are defined. The elasticity of transformation and constant elasticity frontier derived by Powell and Gruen (1968) are discussed. A generalization of the elasticity and the locus are presented, and technical change is discussed in the more general framework.

A STATEMENT OF THE COMMUNITY'S OPTIONS: A TRANSFORMATION CURVE

In the theory of international trade, as well as in the theory of economic exchange, it is essential to define in a consistent manner the opportunity costs of a community's choices of goods and services. The principal analytical tool for such a definition is the production possibility, or transformation, curve.[1] A production function has been defined as the economist's summary of a complex, underlying productive technology for a single good. The transformation curve summarizes the productive technology for two goods with a given resource base, considering all feasible efficient allocations of these resources. It maps the alternative combinations of all the goods and services a community may elect to have, given its resources and state of technology. It does not, however, rely (except for definition of the relevant set of goods and services) upon this preference pattern. Assuming a given resource base and unchanging technology, it is technical data. The curve presents the supply information that a community requires if it is to make efficient decisions.

Before we discuss the derivation of the transformation curve, several assumptions should be explained. First, a single production frontier does not allow for resource increases; the factor inputs are assumed to be both fixed in supply and fully employed. In the construction of the curve, we examine all possible reallocations of these fixed supplies. Second, the productive technology is also given and unchanging. One means of comparing the effects of technical innovations is by contrasting the production possibility curves they imply. Third, we shall assume that factor market imperfections are nonexistent. Factor inputs are allocated as if they were paid the value of their marginal products. If this assumption is relaxed, the shape of the frontier can be distorted so as to prevent the attainment of an equilibrium choice on the part of the community.[2] Finally, we are deliberately ignoring the product market structure. Product market imperfections will prevent the equalization of technical rates of exchange for products on the supply side, with the rates of exchange resulting from the preference patterns on the demand side.

As we noted in the previous chapter, the production possibility frontier can be derived from the underlying production functions, with given assumptions regarding both factor and product markets. Analysis of Baumol's model indicated that his production relationships in each of the two sectors implied a straight-line production possibility curve. The

[1] In what follows we shall use the terms *production possibility curve, transformation curve,* and *production frontier* as synonyms.

[2] There is a growing literature examining the effect of factor market distortions on the production frontier (see Johnson 1966, Jones 1971a, Herberg and Kemp 1971, and Lloyd 1970). If we were to assume that the degree of imperfection were equal across factor markets, then distortions in the shape of the frontier would not prevent an equilibrium. However, the transformation curve in the presence of such distortions would lie within that of perfect markets. This case will be discussed later; for the present we are concerned with a frontier defined as "the best" a community can achieve.

usual shape of this transformation curve is concave to the origin, indicating increasing opportunity costs of production for each commodity. This contour is the result of three factors: (1) the returns to scale in each production activity, (2) the factor intensities (and their possible reversals) in each activity, and (3) the elasticity of substitution between factor inputs in each production function.

Both graphical and algebraic derivations of the production possibility frontier are available in the literature. Most graphical treatments (Melvin 1971b is the leading exception) develop the curve by means of an Edgeworth-Bowley box, plotting the values of the two products involved that correspond to the contract curve tangencies. Thus, for example, in figure 5, OQO' is one contract curve, and along it we can measure the factor input (i.e., X_1 and X_2) utilization and corresponding output levels (i.e., A and B). As we noted, the utilization of X_1 relative to X_2, the factor intensity, in each production process is important to the contour of the production possibility curve. In figure 5, two types of factor intensities are illustrated. Consider a contract curve below a diagonal from O to O'. At the point Q, the factor input ratio (X_1/X_2) in the production of A is OR/OT; that for B is $O'V/O'W$. In this case we can see that A uses more X_2 relative to X_1 than does B. Consequently the factor intensity ratio of B (X_1/X_2) will everywhere along this contract curve exceed that of A. In contrast, the second curve, above the diagonal in figure 5, which represents another mutually exclusive possibility, reverses this ranking. That is, the ratio of X_1 to X_2 in the production of A will everywhere exceed that of B. At point L, it is clear that OG/OE exceeds $O'F/O'D$.

Nothing in the nature of the model suggests that these factor mixes should maintain the same ranking across all ranges of production of A and B. Thus, in some cases, for particular levels of production $(X_1/X_2)_A$ may exceed $(X_1/X_2)_B$; over another range, the opposite will be true. Such changes, called factor intensity reversals, affect the shape of the transformation curve. Figure 6 illustrates one possibility. The ranking at Q corresponds to that of Q in figure 5, as does the relationship of L. In this case, however, there is a single contract curve. Consequently, both rankings can and do occur at different levels of production.

The elasticity of substitution also affects the shape of the transformation curve, since it affects the flatness of any individual isoquant. Accordingly, the magnitude of this elasticity, in absolute as well as relative terms, in the production of A and B will be important.[3]

[3] See Jones (1971a, pp. 448–458) for a discussion of the effect of a *CES* production function on the transformation curve with factor market distortions.

Thus far, we have not specified either the products (A, B) or the inputs (X_1, X_2) used to produce them. Our primary concern is with the role of the services of natural environments in the process of economic growth, and their inherent characteristics may not tolerate the assump-

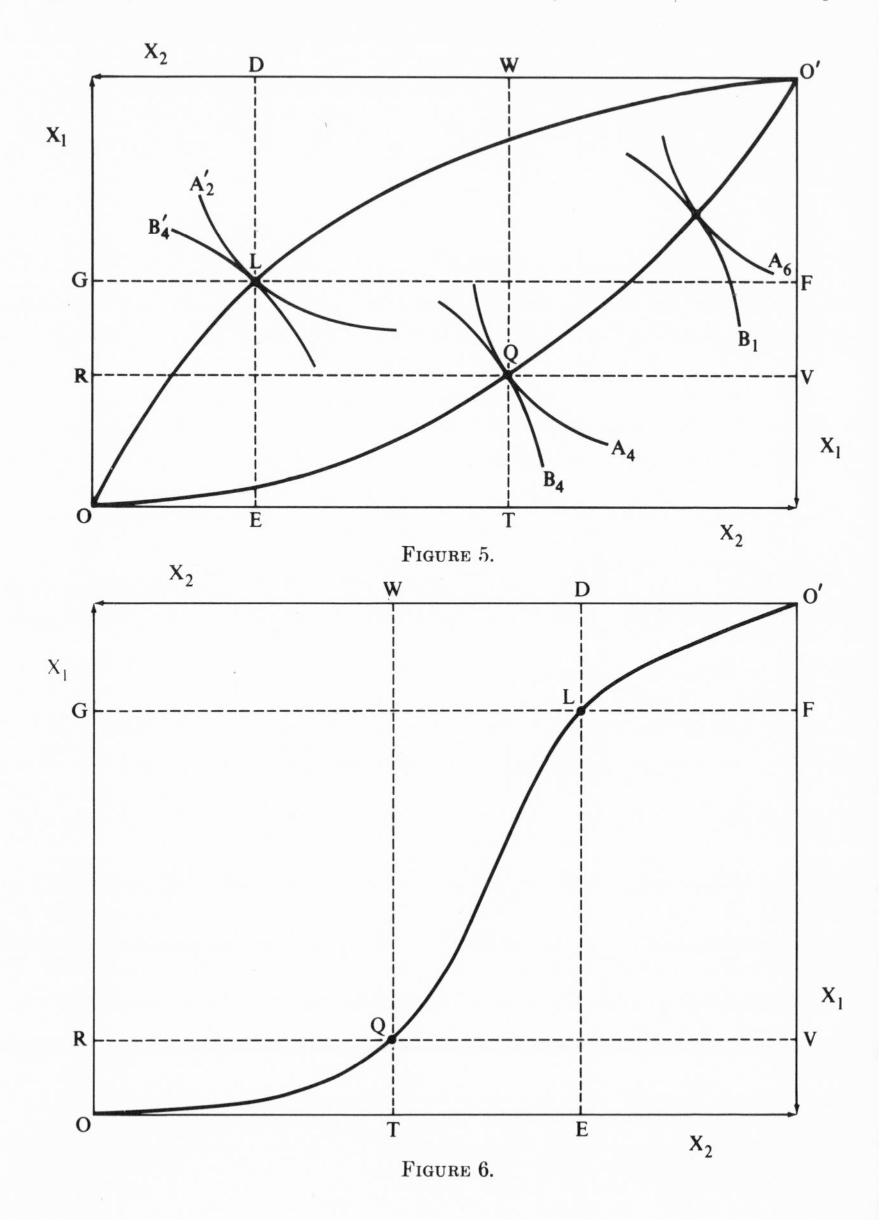

FIGURE 5.

FIGURE 6.

tions we have had to make. This question will be deferred until the next section, in which we shall explore what specifically including environmental resources implies for the transformation curve.

The preceding analysis has focused upon the assumptions necessary to derive a production possibility curve. It is feasible to derive the locus algebraically in simple cases.[4] Assume that the production process for both A and B may be described by using Cobb-Douglas production functions, as in equations (1) and (2):

$$A = aX_{1A}{}^{\alpha}X_{2A}{}^{1-\alpha} \tag{1}$$

$$B = bX_{1B}{}^{\beta}X_{2B}{}^{1-\beta} \tag{2}$$

where: a, b are neutral production parameters
X_{1A} = quantity of X_1 used to produce A
X_{1B} = quantity of X_1 used to produce B
X_{2A} = quantity of X_2 used to produce A
X_{2B} = quantity of X_2 used to produce B
α = elasticity of productivity of X_1 in A
β = elasticity of productivity of X_1 in B

Two important assumptions made in this specification will affect the shape of the resulting transformation curve. First, the elasticity of substitution of X_1 for X_2 in both A's and B's production is unity by specification. In addition, the returns to scale are postulated to be constant by the unitary homogeneity specification. An equally important consequence of these assumptions is the invariance in the relative shares of X_1 and X_2 in each production process.

If we rewrite (1) and (2), making use of the unitary homogeneity of the respective production functions, equations (3) and (4) result:

$$A = aX_{1A}Y_A{}^{1-\alpha} \tag{3}$$

$$B = bX_{1B}Y_B{}^{1-\beta} \tag{4}$$

where: $Y_j = (X_{2_j}/X_{1_j})$; $j = A, B$

Accordingly, Y_A and Y_B are the respective factor mixes in each of the two sectors of the model. Two conditions are required to derive the contract curve and therefore the transformation curve in (A, B) space: (1) all factors must be fully employed and (2) the marginal rate of substitution of X_2 for X_1 in the production of A must equal that for B. Given

[4] This section is based upon a derivation presented by Johnson (1966) and later elaborated upon by Smith and Krutilla (1972).

these, we have equations (5) and (6) describing factor employment (assuming $\bar{X}_1$ and $\bar{X}_2$ are total supplies):

$$\bar{X}_1 = X_{1A} + X_{1B} \tag{5}$$

$$\bar{X}_2 = X_{2A} + X_{2B} \tag{6}$$

It is possible to write these two restrictions in a form that conveys the necessary information.

$$sY_A + (1 - s)Y_B = \bar{Y} \tag{7}$$

where: $s = X_{1A}/\bar{X}_1$

$\bar{Y} = \bar{X}_2/\bar{X}_1$

Y_A and Y_B are as previously defined

Differentiating equations (1) and (2), we can derive the marginal products of X_1 and X_2 in each production process, as in equations (8) through (11). Moreover, with this information it is possible to express the second condition for a contract curve.

$$\frac{\partial A}{\partial X_1} = \alpha a Y_A{}^{1-\alpha} \tag{8}$$

$$\frac{\partial A}{\partial X_2} = (1 - \alpha) a Y_A{}^{-\alpha} \tag{9}$$

$$\frac{\partial B}{\partial X_1} = \beta b Y_B{}^{1-\beta} \tag{10}$$

$$\frac{\partial B}{\partial X_2} = (1 - \beta) b Y_B{}^{1-\beta} \tag{11}$$

For notational convenience A_{X_j} and B_{X_j} will designate the marginal products of factor X_j in the production of A and B respectively. The equilibrium condition that the isoquant slopes be equal gives us (12):

$$\frac{A_{X_1}}{A_{X_2}} = \frac{B_{X_1}}{B_{X_2}} \tag{12}$$

and by direct substitution we have:

$$\frac{\alpha}{1 - \alpha} Y_A = \frac{\beta}{1 - \beta} Y_B \tag{13}$$

The last step in deriving the equivalent of the contract curve is to combine the full employment condition of (7) with the allocative efficiency condition of (13) and find the (X_1, X_2) coordinates that are both exhaustive of the resources and efficient in their use. These are given in (14) and (15):

$$Y_A = \frac{\frac{\beta}{1-\beta}\bar{Y}}{\frac{\alpha}{1-\alpha} + s\left(\frac{\beta-\alpha}{1-\beta-\alpha+\alpha\beta}\right)} \tag{14}$$

$$Y_B = \frac{\frac{\alpha}{1-\alpha}\bar{Y}}{\frac{\alpha}{1-\alpha} + s\left(\frac{\beta-\alpha}{1-\beta-\alpha+\alpha\beta}\right)} \tag{15}$$

In order to simplify the final expression, let:

$$r = \frac{\alpha}{1-\alpha} + s\left(\frac{\beta-\alpha}{1-\beta-\alpha+\alpha\beta}\right)$$

Then equations (14) and (15) may be used to find the corresponding points in the (A, B) space and therefore the transformation locus:

$$\frac{A}{\bar{X}_1} = as\left(\frac{\beta}{1-\beta}\frac{\bar{Y}}{r}\right)^{1-\alpha} \tag{16}$$

$$\frac{B}{\bar{X}_1} = b(1-s)\left(\frac{\alpha}{1-\alpha}\frac{\bar{Y}}{r}\right)^{1-\beta} \tag{17}$$

Alterations in the magnitude of s trace the production possibility frontier.

THE EFFECT OF ENVIRONMENTAL RESOURCES UPON THE TRANSFORMATION LOCUS

Moving from hypothetical models, such as those in the previous section, to more realistic ones, where we designate what actual goods and inputs the symbols stand for, is a difficult and often impossible step. However, if we are to use the models in decision making, then we must examine this transition and be aware of its shortcomings.

Since we are investigating the implications of technical change for environmental resources, our model will be used to make some kind of approximate statement. In so doing we are making a number of restrictive assumptions, postulating a transformation curve concave from the origin and therefore suggesting that the production activities and factor markets associated with environmental resources behave in a particular way. Perhaps, in a somewhat less restrictive mode, we need only to say that the transformation curve between the services of environmental resources and the other good is shaped *as if* the underlying conditions were satisfied. In what follows, this question is addressed in somewhat more detail.

First, the two goods in the model will be designated as manufactured commodities and amenity services derived from environmental resources. The factor inputs necessary to produce each are labor and "direct" services of the environmental resource. It is reasonable to assume that the amenity service good uses much more of the "direct" services of the environmental resource relative to labor than does the manufactured good. But the environmental resources of primary interest are held communally—that is, they are common property resources—and that presents an important problem. Haveman's (1970) definition is perhaps best for our purposes. He notes that "a common property resource is any stock-like 'thing' capable of generating a flow of services, which 'thing' is held communally by a group of individuals" (p. 4). Consequently there is no reason to suspect that the services of the resource will be allocated efficiently between the two production activities.

Conventional economic analysis of the problems associated with common property resources has focused on the effects evidenced from a single firm's behavior or a single industry. In our model, the problem is quite different. The resource can be utilized in either of two ways: to supply amenity services or to produce fabricated goods. In the absence of a functioning market to provide for the allocation of the resource's services, what are the implications for the loci of outputs available to the community as a whole? This question is clearly a difficult one. However, we may be able to approximate the effects by examining a somewhat less extreme case. We shall illustrate the effect of common property resources by considering the externalities they can generate as observationally equivalent to factor market imperfections.

Consider the following substitutions in our algebraic derivation of the previous section:

X_1 = labor
X_2 = direct services of environmental resource
A = amenity services of environmental resource
B = manufactured good

We shall hypothesize that the effect of the common property nature of the direct services of the environment will be to reduce the price of X_2 relative to X_1 in the production of the manufactured good (B) compared with its price in the production of the amenity services. In other words, the price to the producers of the manufactured good does not reflect the marginal social cost of its use.

We can use the model of the previous section to illustrate this condition and to examine the impact of the imperfection upon the shape of the transformation locus. First, in the presence of this imperfection, equation (12) must be rewritten as (18):

$$\frac{A_{X_1}}{A_{X_2}} = \epsilon \frac{B_{X_1}}{B_{X_2}} \tag{18}$$

where: ϵ = the extent of price discrepancy.[5]
(Under our hypothesis, $\epsilon > 0$, and $\epsilon \neq 1$, since X_2 is differentially priced in A and B.)

Equations (14) and (15) describing the optimum factor mix now become (19) and (20):

$$Y_A = \frac{\epsilon\left(\frac{\beta}{1-\beta}\right)\bar{Y}}{\left(\frac{\alpha}{1-\alpha}\right) + s\left(\frac{\beta\epsilon}{1-\beta} - \frac{\alpha}{1-\alpha}\right)} \tag{19}$$

$$Y_B = \frac{\left(\frac{\alpha}{1-\alpha}\right)\bar{Y}}{\left(\frac{\alpha}{1-\alpha}\right) + s\left(\frac{\beta\epsilon}{1-\beta} - \frac{\alpha}{1-\alpha}\right)} \tag{20}$$

Accordingly, the transformation curve may *not* have its concave shape. Consider one possible scenario, $(1-\alpha) > (1-\beta)$. In this case, the utilization of direct services of the environmental resources relative to labor by the manufactured good (B) producers will be greater than that for amenity services. We would have expected the opposite result, and this finding can be interpreted as an indication of how the common property nature of environmental resources encourages their exploitation.

If these findings can be readily generalized, it is clearly reasonable to question the approach selected for modeling and more specifically the use of conventionally shaped production possibility curves. However, there are at least two reasons for continuing with the models. First, it can be argued that we want to bound the possible effects of technical change. That is, even if we forget the effect of the market imperfection, technical change may itself have a serious impact. Thus, introducing the imperfection would further aggravate the situation (assuming the shape

[5] See Johnson (1966, p. 692).

of the locus does *not* prevent equilibrium). Second, and perhaps more important, our present model makes too many restrictive assumptions about the nature of the production process for amenity services and manufactured goods. Neither production process should be represented by a production function with an elasticity of substitution between direct services of the environmental resources and labor that is unity. Given these *a priori* restrictions, an important finding in the derivation of production possibility curves in the presence of factor market imperfections can be utilized. Jones (1971a), using a *CES* formulation, notes that "generally low values of σ_i [elasticity of substitution between two factors in production of ith good] are associated with bowed out transformation schedules" (p. 456) in presence of factor market distortions. Since it is quite likely that both elasticities of substitution will be low, our approximation using a concave production frontier may not be unacceptable. Further reinforcement for our assumptions comes from the work of Scarth and Warne (1973), who suggest that, in the absence of factor market imperfections, low values of one elasticity of substitution will be associated with more bowed-out (concave) transformation curves for a large range of the contract curve (pp. 302–303). In any case, the approach is conservative and will therefore understate the effects of technology (at least in their supply influence) rather than overstate them.

AUTONOMOUS TECHNICAL CHANGE AND THE TRANSFORMATION CURVE

Autonomous technical change may be defined as change in the productive technology that is determined outside the economic model under study. Thus both the magnitude and the direction of the change are predetermined and beyond the influence of the economy. Most of the literature dealing with technological innovation has specified the effect of such changes upon the production functions. They are assumed neutral or non-neutral with respect to the factors. As chapter 2 suggests, it is more useful for our purposes to redefine the effect of the change in terms of its effects upon products. Given a mutually exclusive and exhaustive list of the goods and services a community might desire and a given stock of input resources, we can examine the effects of technological changes upon the goods and services. For example, in a two-commodity world, a neutral change affects both goods equally. In figure 7 the original transformation curve is SS'; after a neutral change it becomes TT'. The locus shifts out in a parallel fashion. By contrast, a non-neutral change has a greater effect upon one of the two goods. Thus in figure 8 the shift from SS' to ST is a non-neutral change. For every possible amount of A, the quantity of B has been increased.

Both of these cases can be illustrated in the Edgeworth-Bowley (E-B) box framework. However, different technical changes, in terms of the production functions and factor inputs, can yield equivalent changes in the production possibility curve. For example, if both production functions (A's and B's) are homogeneous of degree one, then a neutral change in terms of the products can be seen as an equiproportionate increase

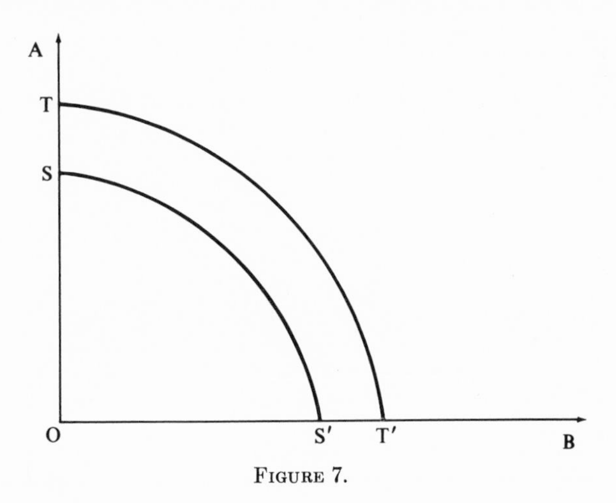

FIGURE 7.

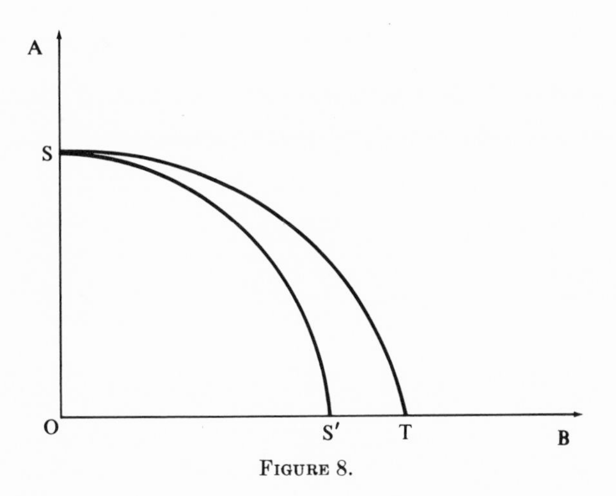

FIGURE 8.

in the size of the E-B box.[6] This situation is illustrated in figure 9. Movement from the O_0O_0' E-B box to the O_1O_1' is equivalent to movement from SS' to TT' in figure 7.

Though it is conceivable, it is not possible, without explicit specification of the production functions in question, to prove that a non-neutral change in terms of factors for the production of B with no alteration in the production process for A (e.g., a change to dashed isoquants in figure 10) will imply a shift in the transformation curve from SS' to ST in figure 8. Accordingly the important points to be made here are that factor-neutral and non-neutral technical changes are not synonymous with product-neutral changes. Moreover, a number of alternative innovation patterns in terms of factor neutrality will yield equivalent results in the product space. Our primary concern, therefore, will be to define the changes in terms of how they affect the final outputs rather than the production functions themselves.

IMPLICATIONS OF A *CET* TRANSFORMATION FRONTIER

A production function has been described as the economist's summary of a complex underlying productive technology. In order to increase the understanding of that relationship, certain characteristics of the function were also described. It is convenient for the present analysis to postulate a production transformation curve and to assume that the underlying production functions and factor markets yield the function. This form of analysis does not differ in principle from postulating that a simple relationship between inputs and outputs "explains" a complex underlying system. Accordingly, several parameters of the transformation locus will also be presented as descriptive of the locus and the underlying production functions and factor markets.

The first of these, the elasticity of transformation, measures the curvature of a production possibility curve. It is negative because of the concave shape of the production frontier, and similar in definition to the elasticity of substitution between factors. In this case, however, we are measuring the percentage change in the product mix that accompanies a percentage change in the relative opportunity costs. In figure 11, it is

[6] Suppose we have two production functions: $f_A(x_1, x_2)$ and $f_B(x_1, x_2)$. If they are homogeneous of degree 1, equal proportionate increases in x_1 and x_2 are the same as a neutral change:

$$\lambda f_A(x_1, x_2) = f_A(\lambda x_1, \lambda x_2) \text{ and } \lambda f_B(x_1, x_2) = f_B(\lambda x_1, \lambda x_2)$$

represented by the percent change in the slope of OP relative to that of RR'. We can write it thus:

$$\tau = \frac{\dfrac{d\,(A/B)}{A/B}}{\dfrac{d\,(\partial A/\partial B)}{\partial A/\partial B}}; \; \tau < 0 \tag{21}$$

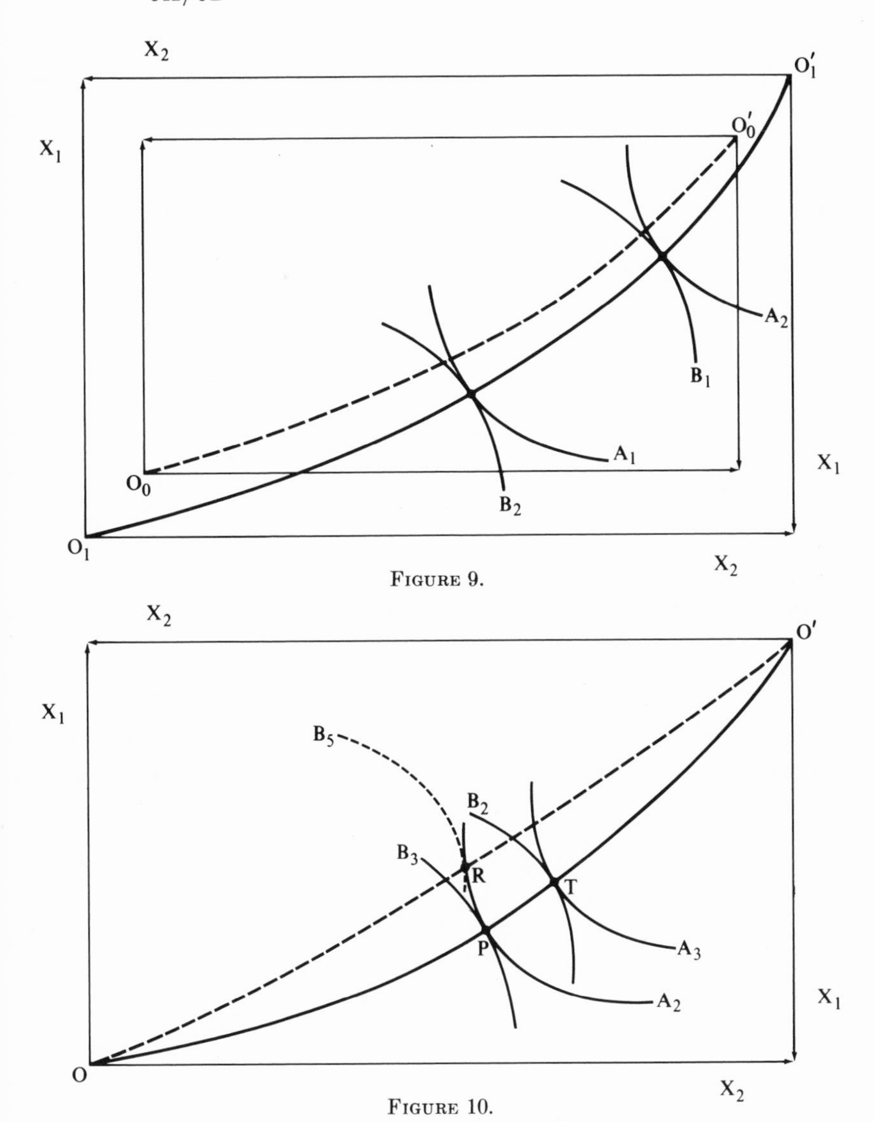

FIGURE 9.

FIGURE 10.

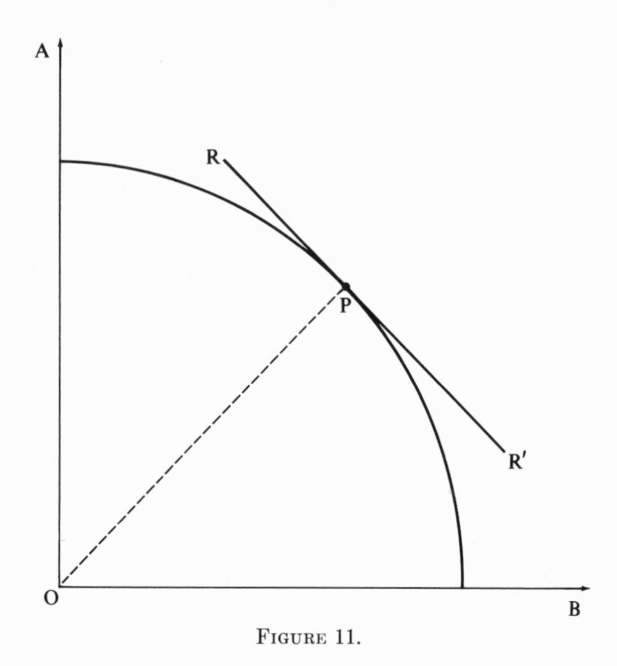

FIGURE 11.

Powell and Gruen (1968) introduced this index and derived a constant elasticity of transformation (*CET*) locus, in a spirit similar to the constant elasticity of substitution (*CES*) production function.[7] Our attention will focus upon *CET* frontiers. Equation (22) describes a two-good *CET* locus:

$$\frac{1}{\alpha} B^{1-1/\tau} + A^{1-1/\tau} = R(1 - 1/\tau) \tag{22}$$

Changes in R represent neutral shifts in the production locus, as in figure 7, whereas an increase in α represents a non-neutral movement in the locus, as represented in figure 8 with the transformation curve moving from SS' to ST. As the absolute value of the elasticity τ increases, the frontier approaches a downward-sloping straight line.

If we assume that factor supplies are unchanging, then movements in the transformation curve can be assumed to reflect changes in the technology. Moreover, the previously defined concepts of neutral and non-neutral changes can be linked to two parameters of the function R and α, respectively.

[7] Further generalization of the Powell-Gruen frontier has been developed by Christensen, Jorgenson, and Lau (1973).

In order to examine a model with more than two goods, we must generalize the elasticity of transformation to a partial elasticity framework. Equation (23) defines the partial elasticity:[8]

$$\tau_{ij} = \frac{Z_1 g_1 + Z_2 g_2 + \cdots + Z_n g_n}{Z_i Z_j} \cdot \frac{H_{ij}}{H} \tag{23}$$

where: $Z_1, \cdots, Z_n$ = commodities

$g(Z_1, Z_2, \cdots, Z_n)$ = production transformation locus

$$g_i = \frac{\partial g}{\partial Z_i}$$

$$g_{ij} = \frac{\partial^2 g}{\partial Z_j \partial Z_i}$$

$$H = \det \begin{bmatrix} 0 & g_1 & g_2 & \cdots & g_n \\ g_1 & g_{11} & g_{12} & \cdots & g_{1n} \\ g_2 & g_{21} & g_{22} & \cdots & g_{2n} \\ \cdot & \cdot & & & \cdot \\ \cdot & \cdot & & & \cdot \\ \cdot & \cdot & & & \cdot \\ g_n & g_{n1} & g_{n2} & \cdots & g_{nn} \end{bmatrix}$$

H_{ij} = determinant of the ij^{th} cofactor of the bordered Hessian above

$\tau_{ij} < 0$

This definition is analogous to the Uzawa (1962) partial elasticity of substitution. Moreover, the partial *CET* frontier is similar in form and given in equation (24):

$$\alpha_1 Z_1^{1-1/\tau} + \alpha_2 Z_2^{1-1/\tau} + \alpha_3 Z_3^{1-1/\tau} + \cdots + \alpha_n Z_1^{1-1/\tau} = K'(1 - 1/\tau) \tag{24}$$

where: $\tau_{12} = \tau_{13} = \cdots = \tau_{21} = \tau_{23} \cdots = \tau_{n1} = \tau$

This can be demonstrated fairly simply in the three-commodity case, and the n-good case is a straightforward generalization. Suppose a transformation curve is given in (25):

$$\alpha_1 Z_1^{-B} + \alpha_2 Z_2^{-B} + \alpha_3 Z_3^{-B} = A(1 - 1/\tau) = g(Z_1, Z_2, Z_3) \tag{25}$$

then: $g_{12} = g_{21} = g_{31} = g_{13} = g_{23} = g_{32} = 0$

[8] Powell (1971) and Smith (1972) developed analogous definitions of this generalized elasticity independently. It is defined in the same spirit as Uzawa's (1962) partial elasticity of substitution.

To derive the elasticity of transformation, we need only substitute into (23), and we get for H and H_{12} the expressions in (26a) and (26b), respectively:

$$H = \det\begin{bmatrix} 0 & -\alpha_1 BZ_1^{-B-1} & -\alpha_2 BZ_2^{-B-1} & -\alpha_3 BZ_3^{-B-1} \\ -\alpha_1 BZ_1^{-B-1} & \alpha_1(B+1)Z_1^{-B-2} & 0 & 0 \\ -\alpha_2 BZ_2^{-B-1} & 0 & \alpha_2 B(B+1)Z_2^{-B-2} & 0 \\ -\alpha_3 BZ_3^{-B-1} & 0 & 0 & \alpha_3 B(B+1)Z_3^{-B-2} \end{bmatrix} \tag{26a}$$

$$H_{12} = \det\begin{bmatrix} 0 & -\alpha_2 BZ_2^{-B-1} & -\alpha_3 BZ_3^{-B-1} \\ -\alpha_1 BZ_1^{-B-1} & 0 & 0 \\ -\alpha_3 BZ_3^{-B-1} & 0 & \alpha_3 B(B+1)Z_3^{-B-2} \end{bmatrix} \tag{26b}$$

So we have H and H_{12} in more simple terms as (27a) and (27b). Substitution into the remainder of (23) yields the expression for τ_{12} in (28).

$$\begin{aligned} H = &-\alpha_1^2\alpha_2\alpha_3 B^4(B+1)^2 Z_1^{-2B-2}Z_2^{-B-2}Z_3^{-B-2} \\ &-\alpha_1\alpha_3^2\alpha_2 B^4(B+1)^2 Z_1^{-B-2}Z_2^{-B-2}Z_3^{-2B-2} \\ &-\alpha_1\alpha_2^2\alpha_3 B^4(B+1)^2 Z_1^{-B-2}Z_2^{-2B-2}Z_3^{-B-2} \end{aligned} \tag{27a}$$

$$H_{12} = \alpha_1\alpha_2\alpha_3 B^3(B+1)Z_1^{-B-1}Z_2^{-B-1}Z_3^{-B-2} \tag{27b}$$

$$\tau_{12} = 1/(1+B) \tag{28}$$

Since B is a constant, the partial elasticities are all constant and independent of product prices. It can be demonstrated in an analogous fashion that $\tau_{13} = \tau_{12} = \tau_{23} = 1/(1+B)$.

Once the model is generalized to include more than two goods, specifying non-neutral change becomes somewhat more complicated. The relative changes in α_1, α_2, and α_3 become important. For example, suppose that we add to our two-commodity model another good—call it C; then the transformation locus might be given as:

$$\alpha_1 A^{1-1/\tau} + \alpha_2 B^{1-1/\tau} + \alpha_3 C^{1-1/\tau} = K'(1 - 1/\tau) \tag{29}$$

Changes in K' will result in parallel shifts in the transformation surface. However, non-neutral technical changes must be specified in terms of the relative shifts of each intercept of the locus. For example, in figure 12 the transformation contour is given by SRS'. A technical innovation might shift the frontier in the next period to SDT. Such a change is non-neutral. Moreover, the increase along the B axis has been greatest, since $S'T$ exceeds RD and the A coordinate has not altered. Analytically, specification of such changes requires a statement of how α_2/α_1 is changing relative to α_3/α_1. If α_2/α_1 is increasing faster than α_3/α_1, then technical change would be pushing the A coordinate out the most, C next,

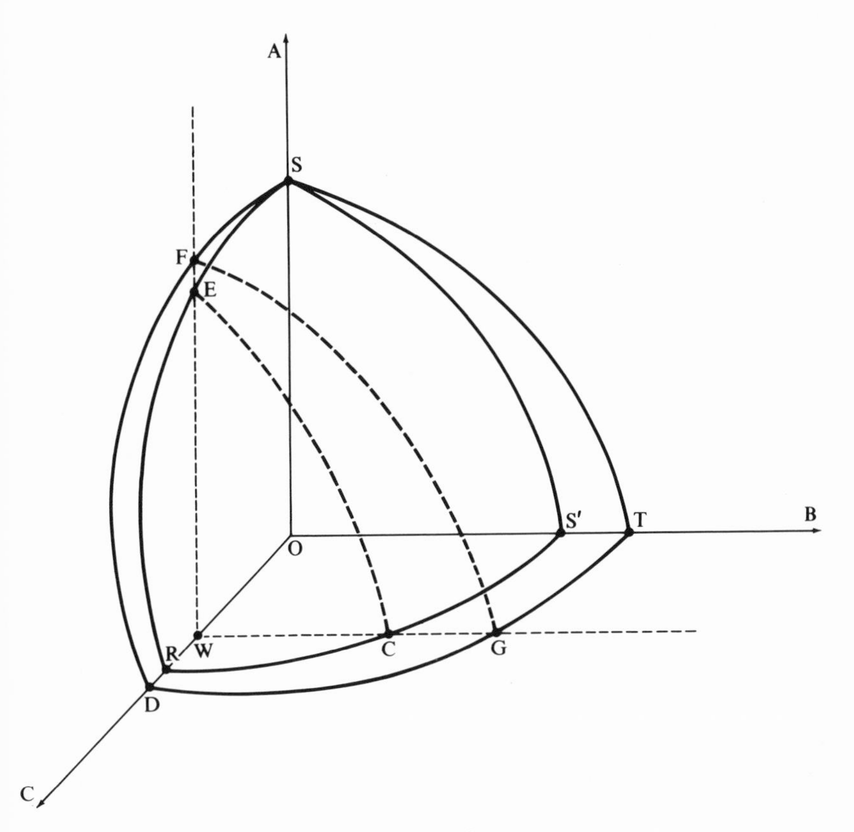

FIGURE 12.

and B least. If we have α_2/α_1 declining more than α_3/α_1 (which is also declining), then the pattern from SRS' to SDT is the result for the transformation surface.

Further discussion will be delayed until the complete model is treated in chapter 5. It should be noted, however, that a given cross-section of the transformation surface may disguise the overall effect of technology in terms of products. If, for example, we were to examine the cross-section in the CB plane with A held at zero, the transformation curve shifts from ORS' to ODT; while there is non-neutral progress the quantities of both goods are increasing. Alternatively, holding either B or C at zero and examining the other relative to A indicates the increasing relative scarcity of A as a result of technology. Finally, holding C at OW, we have yet another comparison, WCE relative to WGF. The primary point is simply that partial pictures of the effect of technical innovation can be deceiving. This finding also holds true for product-related definitions of technical change.

SUMMARY

Our review of the transformation locus, which is a technical description of the efficient combinations of goods and services available to a community with a given resource base, has raised several points.

(1) A production possibility curve maps out the commodity combinations available with allocations of factor inputs, wherein the marginal contributions of each are equalized in each activity.

(2) The transformation locus may be derived using an Edgeworth-Bowley box. Its concave shape reflects the increasing opportunity costs that increased production of a commodity entails.

(3) Three factors affect the traditional shape of the curve: (a) the returns to scale in each production activity; (b) the factor intensities in each activity as well as reversals of these intensities under different output levels; and (c) the elasticity of substitution between the factor inputs used in each production activity.

(4) If we assume that the direct services of an environmental resource are one factor input, then we introduce a market imperfection into the factor markets. With Cobb-Douglas production functions, the contour may be convex. However, if the elasticities of substitution between factor inputs in each production process are small, we may expect the usual shape of the contour to be retained in spite of the imperfection.

(5) Neutral technical change, defined in terms of products, means the transformation curve has shifted out in a parallel fashion. Nonneutral changes imply nonparallel shifts.

(6) The elasticity of transformation measures the curvature of the production possibility curve and is defined as the percent change in the product mix relative to the percent change in the marginal rate of transformation.

REFERENCES

Batra, R., and P. K. Pattanaik. 1971. "Economic Growth, Intermediate Products, and the Terms of Trade." *Canadian Journal of Economics* 4: 225–237.

Bhagwati, J., and V. K. Ramaswami. 1963. "Domestic Distortions, Tariffs, and the Theory of Optimum Subsidy." *Journal of Political Economy* 71: 44–50.

Black, J. 1957. "A Formal Proof of the Concavity of the Production Possibility Function." *Economic Journal* 67: 133–135.

Christensen, L. R., D. W. Jorgenson, and F. J. Lau. 1973. "Transcendental Logarithmic Production Frontiers." *Review of Economics and Statistics* 55: 28–45.

Corden, W. M. 1956. "Economic Expansion and International Trade: A Geometric Approach." *Oxford Economic Papers* 8: 223–228.

Ferguson, C. E. 1962. "Transformation Curve in Production Theory: A Pedagogical Note." *Southern Economic Journal* 29: 96–102.

———. 1969. *Microeconomic Theory*. Homewood, Ill.: Richard D. Irwin, Inc.

Gruen, F. H. 1961. "Agriculture and Technical Change." *Journal of Farm Economics* 43: 838–858.

Haberler, G. 1961. *A Survey of International Trade Theory*. International Finance Section, Princeton University.

Haveman, R. 1970. "The Economics of Common Property Resources." Unpublished paper.

Heller, H. R. 1968. *International Trade: Theory and Empirical Evidence*. Englewood Cliffs, N.J.: Prentice-Hall, Inc.

Herberg, H., and M. C. Kemp. 1971. "Factor Market Distortions, the Shape of the Locus of Competitive Outputs, and the Relation between Output Prices and Equilibrium Outputs." In *Trade, Balance of Payments, and Growth*, ed. J. N. Bhagwati et al. Amsterdam: North Holland Publishing Co.

Inada, Ken-Ichi. 1965. "On Neoclassical Models of Economic Growth." *Review of Economic Studies* 32: 151–160.

Johnson, H. G. 1966. "Factor Market Distortions and the Shape of the Transformation Curve." *Econometrica* 34(3): 686–698.

Jones, R. W. 1965. "The Structure of Simple General Equilibrium Models." *Journal of Political Economy* 73: 557–572.

———. 1971a. "Distortions in Factor Markets and the General Equilibrium Model of Production." *Journal of Political Economy* 79: 437–459.

———. 1971b. "A Three Factor Model in Theory, Trade, and History." In *Trade, Balance of Payments, and Growth*, ed. J. N. Bhagwati et al. Amsterdam: North Holland Publishing Co.

Kelly, J. S. 1969. "Lancaster vs. Samuelson on the Shape of the Neoclassical Transformation Surface." *Journal of Economic Theory* 1: 347–351.

Lancaster, K. 1968. *Mathematical Economics*. New York: Macmillan Co.

Leontief, W. W. 1933. "The Use of Indifference Curves in the Analysis of Foreign Trade." *Quarterly Journal of Economics* 47: 493–503.

Lerner, A. P. 1936. "The Symmetry between Import and Export Taxes." *Economica* 3: 306–314.

Lloyd, P. J. 1970. "The Shape of the Transformation Curve with and without Factor Market Distortions." *Australian Economic Papers* 9: 52–62.

Melvin, J. R. 1968. "Production and Trade with Two Factors and Three Goods." *American Economic Review* 58(5): 1249–68.

———. 1971a. "International Trade Theory without Homogeneity." *Quarterly Journal of Economics* 85: 66–76.

———. 1971b. "On the Derivation of the Production Possibility Curve." *Economica* 38: 287–294.

Powell, A. A. 1971. *Informative Estimation of Economic Models: The Empirical Analytics of Demand and Supply.* Unpublished monograph, Monash University.

Powell, A. A., and F. H. Gruen. 1968. "The Constant Elasticity of Transformation Production Frontier and Linear Supply System." *International Economic Review* 9(3): 315–328.

Savosnick, K. M. 1958. "The Box Diagram and the Production Possibility Frontier." *Ekonomisk Tidsskrift* 60: 183–197.

Scarth, W. M., and R. D. Warne. 1973. "The Elasticity of Substitution and the Shape of the Transformation Curve." *Economica* 40: 299–304.

Scitovszky, T. 1942. "A Reconsideration of the Theory of Tariffs." *Review of Economic Studies* 9: 89–110.

Smith, V. K. 1972. "The Incidence of Technological Change among Different Uses of Environmental Resources." In *Natural Environments: Studies in Theoretical and Applied Analysis*, ed. J. V. Krutilla. Baltimore: The Johns Hopkins University Press for Resources for the Future, Inc.

Smith, V. K., and J. V. Krutilla. 1972. "Technical Change and Environmental Resources." *Socio-Economic Planning Science* 6: 125–132.

Stewart, D. B. 1971. "Production Indeterminacy with Three Goods and Two Factors: A Comment on the Pattern of Trade." *American Economic Review* 61(1): 241–244.

Stopler, W. F., and P. A. Samuelson. 1941. "Protection and Real Wages." *Review of Economic Studies* 8: 58–73.

Uzawa, H. 1962. "Production Functions with Constant Elasticities of Substitution." *Review of Economic Studies* 29: 291–299.

Vanek, J. 1963. "Variable Factor Proportions and Inter-Industry Flows in the Theory of International Trade." *Quarterly Journal of Economics* 77: 129–142.

Worswick, D. 1957. "The Convexity of the Production Possibility Function." *Economic Journal* 67: 748–750.

chapter four The Structure of Community Demand

We shall model the demand side of a general equilibrium model through the specification of a community utility function. Since the commodities and services included in the model contain the amenity services provided by environmental resources, some consideration is given the nature of the demand for such services. Moreover, this information must be used in discriminating between alternative possible utility specifications and their resulting demand structures. The Hicks-Allen equations relating utility functions to demand structure provide this information and a nonhomogeneous function proves to be the most flexible.

THE NATURE OF THE DEMAND FOR AMENITY SERVICES

Before outlining the criteria that must be imposed upon any model of community demand, we must examine the empirical evidence from partial equilibrium models that is related to the demand characteristics of the goods and services we are considering. Special attention will be given to the demand for amenity services derived from preserved natural endowments. One of the most important members of this class of services is outdoor recreation.

Over the past decade, the demand for outdoor recreational experiences has received the attention of an increasing number of economists. Although the results of their analyses have not always been definitive, several observations are possible. Cicchetti, Seneca, and Davidson (1969) have found that tastes for much outdoor recreation, particularly wilderness recreation, may change over time and affect willingness to pay for the experiences. Two factors affect growth in demand for these services. First, as incomes increase over time, the luxury character (i.e., with income elasticities of demand greater than unity) of outdoor recreational activities implies greater participation. Second, participation itself is a learning experience that reinforces and increases demand, without diminishing present supply (see Davidson, Adams, and Seneca 1966, p. 186). In a study of wilderness recreation, Cicchetti (1972) finds that the age at which an individual first encounters wilderness recreation also influences the probability of his continued participation.

The availability of recreational areas certainly affects present use. In addition, intertemporal reinforcement exerted by present use on future demand through the *learning by doing* phenomenon links the availability of present supply with the level of future demand. Furthermore, the services of wilderness recreation have few adequate substitutes; their unique attributes are not supplied by other forms of recreation (see Stankey 1972).

Any model of community preferences must be able to accommodate a fairly diverse demand structure. That is, those goods which form the rest of the community's desires are not likely to have the same pattern of price, cross, and income elasticities as amenity services do. Many of these goods will be substitutes for each other, and some may be insensitive to income change. Accordingly, the mode chosen for representing community preferences must be sufficiently flexible to reflect some of these demand patterns.

There is an additional issue involved with modeling such demand patterns. Typically, the amenity services derived from a natural endowment are provided in an "extra-market" fashion. Thus the user fees paid for such services reflect neither the costs of provision nor the users' willingness to pay (see Cicchetti et al. 1972). Estimating the demand for such services is an empirical exercise, focusing on two problems: (1) Is it possible to identify a structural demand equation with the available information? (2) Can we measure partial equilibrium demand functions for a consistently defined amenity service holding constant the nonmonetary costs involved in its consumption? Both questions are beyond the scope of our analysis (see Cicchetti, Fisher, and Smith 1973); the demand structure of the community is *assumed* known for the present purposes. Consequently, it is possible to discern the effects of such exogenous changes as technical advance and intertemporal production externalities upon the community's choices. One might argue that since the services of the endowments are not exchanged in well-defined markets, price is a misnomer. To avoid this problem, relative prices will be used in the following analysis to reflect the community's relative valuations of its goods and services.

INDIVIDUAL VERSUS COMMUNITY UTILITY FUNCTIONS

Most analyses of demand are based on an individual's utility function, so that application to community demand necessarily requires specification of a community utility function.[1] Much of the extensive literature on the use of indifference curves for measurement of community welfare is associated with international trade theory. In a careful study of these analytical tools, Samuelson (1956) notes that "community indifference curves between the totals of two goods . . . give us a 'demand relationship.' . . . *and essentially nothing more*" (p. 4). While community indifference curves are an alternative means of describing a community's demand structure, they do not provide a viable means of determining the relationship between individual utility and community welfare. Consequently, attempts to derive them through a series of restrictive aggregation schemes result in community indifference curves that must have everyone in the society alike in all respects,[2] an assumption untenable for most analyses. But we do not have to make this assumption in order to use a community utility function. We need only follow Samuelson's advice and view the function as an alternative means for describing group demand, without considering the individual in relation to that group.

A second approach, observationally equivalent in its results, is based on "the myth of the representative consumer, which underlies the application of a theory based on individual consumer behavior" (Parks 1969, p. 630, n. 4) to an aggregate problem. The assumption that the com-

[1] Strictly speaking, the manner in which such a utility function is specified can be changed in either of the approaches to accommodate other than a cardinal relationship. See Lancaster (1968), Katzner (1970), and Hicks and Allen (1934).

[2] See Heller (1968, pp. 45–51) for a summary of these assumptions.

munity behavior can be explained by reference to the representative consumer, characteristic of all studies of community expenditure patterns, conveniently avoids the difficulties associated with the aggregation problem (see Powell 1971, pp. 65–66). If the group's choices can be represented by one person, we need only focus attention on a utility function that would describe his behavior. Our analysis of the price behavior of amenity services relative to that of fabricated goods would thus be for this representative individual.

Accordingly, whether we choose the first or the second rationale, we shall be working with a specification of community utility, recognizing that this is simply an alternative means of describing a preassigned pattern of demand for the group.

THE HICKS-ALLEN EQUATIONS

The Hicks-Allen (1934) paper presents the interrelationships between the parameters describing an individual's demand structure in the perfectly general case as well as under the assumption of integrability (i.e., the existence of an integral of the fundamental differential equation). Most treatments of individual demand theory have followed the Slutsky tradition and derive the Slutsky equation with its income and substitution effects.[3] Without the benefit of his work, Hicks and Allen developed a completely analogous description of the individual's behavior in terms of elasticities. Unfortunately this work has received less attention than that of Slutsky. In what follows, the Hicks-Allen equations will be derived for the two-commodity case, and then the more general equations for the three-good (or more) case will be reviewed.

The first order conditions for utility maximization result in two conditions that must be satisfied as in (1) and (2) below.

$$X_1P_1 + X_2P_2 = y \tag{1}$$

$$R_{X_1}{}^{X_2} = \frac{P_2}{P_1} \tag{2}$$

where: X_1 = quantity of good 1

X_2 = quantity of good 2

P_1 = price of X_1

P_2 = price of X_2

y = money income

$R_{X_1}{}^{X_2} = -\dfrac{dX_1}{dX_2}$ (marginal rate of substitution between X_1 and X_2)

[3] See Slutsky (1915), and also Lancaster (1968) for a complete discussion of the Slutsky equations.

In order to analyze individual behavior, Hicks and Allen ask how the individual will react if P_1 changes but he must maintain (1) and (2) in adjusting to the price change. This behavior can be examined by partially differentiating (1) and (2) with respect to P_1. However, before presenting these partial differential equations, some definitional relationships should be discussed. Hicks and Allen define coefficients of income variation as in (3a) and (3b) below.

$$\rho_{X_1} = -\frac{X_2 P_1}{P_2} \cdot \frac{\partial}{\partial X_2} (R_{X_1}{}^{X_2}) \tag{3a}$$

$$\rho_{X_2} = \frac{X_1 P_1}{P_2} \cdot \frac{\partial}{\partial X_1} (R_{X_1}{}^{X_2}) \tag{3b}$$

The elasticity of substitution between X_1 and X_2, σ, may be defined along a given indifference curve as:

$$\sigma = -\frac{\dfrac{y}{X_1 X_2} \cdot \dfrac{P_2}{P_1}}{P_1 \dfrac{\partial}{\partial_{X_2}} (R_{X_1}{}^{X_2}) - P_2 \dfrac{\partial}{\partial_{X_1}} (R_{X_1}{}^{X_2})} \tag{4}$$

Given budget exhaustion, the income elasticities of demand for X_1 and X_2, defined as $E_y(X_1)$ and $E_y(X_2)$, respectively, are related to the coefficients of income variation and the elasticity of substitution by the following relationships (see Hicks and Allen 1934, pp. 200–201):

$$E_y(X_1) = \frac{\partial X_1}{\partial y} \cdot \frac{y}{X_1} = \rho_{X_1} \cdot \sigma \tag{5}$$

$$E_y(X_2) = \frac{\partial X_2}{\partial y} \cdot \frac{y}{X_2} = \rho_{X_2} \cdot \sigma \tag{6}$$

With this equipment it is possible to relate the own price and cross elasticities of demand to the income elasticity of demand and the elasticity of substitution. Partially differentiating (1) with respect to P_1, we have:

$$X_1 + P_1 \frac{\partial X_1}{\partial P_1} + P_2 \frac{\partial X_2}{\partial P_1} = 0 \tag{7}$$

Some algebraic manipulation allows (7) to be rewritten as:

$$P_1 X_1 = P_1 X_1 \left(-\frac{P_1}{X_1} \frac{\partial X_1}{\partial P_1} \right) + P_2 X_2 \left(-\frac{P_1}{X_2} \frac{\partial X_2}{\partial P_1} \right) \tag{8}$$

In similar fashion, if we differentiate (2) with respect to P_1, (9) results:

$$R_{X_1}{}^{X_2} + P_1 \frac{\partial}{\partial X_1} (R_{X_1}{}^{X_2}) \frac{\partial X_1}{\partial P_1} + P_1 \frac{\partial}{\partial X_2} (R_{X_1}{}^{X_2}) \frac{\partial X_2}{\partial P_1} = 0 \tag{9}$$

Substituting for $R_{X_1}{}^{X_2}$ from (2) and simplifying, we have:

$$\frac{P_2}{P_1} = X_1 \frac{\partial}{\partial X_1}(R_{X_1}{}^{X_2})\left(-\frac{P_1}{X_1}\frac{\partial X_1}{\partial P_1}\right) + X_2 \frac{\partial}{\partial X_2}(R_{X_1}{}^{X_2})\left(-\frac{P_1}{X_2}\frac{\partial X_2}{\partial P_1}\right) \quad (10)$$

If we define the price elasticities of demand for X_1 as in (11a) and (11b), then (8) and (10) may be rewritten as (12) and (13):

$$E_{P_1}(X_1) = -\frac{P_1}{X_1}\frac{\partial X_1}{\partial P_1} \quad (11a)$$

$$E_{P_1}(X_2) = -\frac{P_1}{X_2}\frac{\partial X_2}{\partial P_1} \quad (11b)$$

$$P_1X_1 = P_1X_1E_{P_1}(X_1) + P_2X_2E_{P_1}(X_2) \quad (12)$$

$$\frac{P_2}{P_1} = X_1 \frac{\partial}{\partial X_1}(R_{X_1}{}^{X_2})E_{P_1}(X_1) + X_2 \frac{\partial}{\partial X_2}(R_{X_1}{}^{X_2})E_{P_1}(X_2) \quad (13)$$

Utilizing the definitions in (3a), (3b), (5), and (6), we have substitutes for the partial derivatives of the marginal rate of substitution in (13) as follows:

$$\frac{P_2}{P_1} = \frac{P_2}{P_1}\frac{E_y(X_2)}{\sigma}E_{P_1}(X_1) - \frac{P_2}{P_1}\frac{E_y(X_1)}{\sigma}E_{P_1}(X_2) \quad (14)$$

Algebraic simplification yields:

$$\sigma = E_y(X_2)E_{P_1}(X_1) - E_y(X_1)E_{P_1}(X_2) \quad (15)$$

Substituting from (12) for $E_{P_1}(X_1)$, we have:

$$E_{P_1}(X_2) = \frac{P_1X_1}{y} \cdot E_y(X_2) - \frac{P_1X_1}{y}\sigma \quad (16)$$

In similar fashion, equation (17) for $E_{P_1}(X_1)$ can be derived from (15) and (12):

$$E_{P_1}(X_1) = \frac{P_1X_1}{y} \cdot E_y(X_1) + \frac{P_2X_2}{y}\sigma \quad (17)$$

These equations should be interpreted as alternatives to the more conventional Slutsky equation. In the more general case, with three goods (X_1, X_2, and X_3) the derivation follows the same general outline and will therefore not be presented.

When the integrability condition is satisfied (requiring more than the weak axiom of revealed preference), then a complete indifference surface will exist in (X_1, X_2, X_3) space. Hicks and Allen assume that the marginal rates of substitution—defined in (18) and (19) if $f(X_1, X_2, X_3)$ is the utility function—are continuous functions of X_1, X_2, and X_3. Moreover, they are positive at all points. Finally, for a movement along the indifference surface, the fundamental differential equation, defined in (20), describes the compensating changes in the marginal utility of each good.

$$R_{X_1}{}^{X_2} = -\frac{dX_1}{dX_2} = \frac{f_{X_2}}{f_{X_1}} \tag{18}$$

$$R_{X_1}{}^{X_3} = -\frac{dX_1}{dX_3} = \frac{f_{X_3}}{f_{X_1}} \tag{19}$$

$$dX_1 + R_{X_1}{}^{X_2}dX_2 + R_{X_1}{}^{X_3}dX_3 = 0 \tag{20}$$

The preceding assumptions assure the downward slope of the indifference surface in either of the planes in which a cross section might be taken. Moreover, the second assures the increasing marginal rate of substitution and usual shape of the indifference surface.

Before describing the equations linking demand parameters we must set out some definitions. First, some measure of the curvature of the indifference surface passing through a point (σ^*, defined in equation [21]) must be defined. This index measures the mutual substitutability among the three goods. Second, a measure of the paired associations is also important. For this comparison we define the partial elasticity of substitution in equation (22).

$$\sigma^* = \frac{\frac{dX_1}{dX_2} \cdot \frac{dX_1}{dX_3}}{X_1 \cdot X_2 \cdot X_3} \cdot \frac{X_1 - \frac{dX_1}{dX_2} \cdot X_2 - \frac{dX_1}{dX_3} \cdot X_3}{G} \tag{21}$$

$$G = \det \begin{bmatrix} 1 & -\frac{dX_1}{dX_2} & -\frac{dX_1}{dX_3} \\ \frac{\partial}{\partial X_1}\left(-\frac{dX_1}{dX_2}\right) & \frac{\partial}{\partial X_2}\left(-\frac{dX_1}{dX_2}\right) & \frac{\partial}{\partial X_3}\left(-\frac{dX_1}{dX_2}\right) \\ \frac{\partial}{\partial X_1}\left(-\frac{dX_1}{dX_3}\right) & \frac{\partial}{\partial X_2}\left(-\frac{dX_1}{dX_3}\right) & \frac{\partial}{\partial X_3}\left(-\frac{dX_1}{dX_3}\right) \end{bmatrix}$$

$$\sigma_{ij} = \frac{X_1 f_{X_1} + X_2 f_{X_2} + X_3 f_{X_3}}{X_i X_j} \cdot \frac{F_{ij}}{F} \tag{22}$$

where:

$$F = \det \begin{bmatrix} 0 & f_{X_1} & f_{X_2} & f_{X_3} \\ f_{X_1} & f_{X_1X_1} & f_{X_1X_2} & f_{X_1X_3} \\ f_{X_2} & f_{X_2X_1} & f_{X_2X_2} & f_{X_2X_3} \\ f_{X_3} & f_{X_3X_1} & f_{X_3X_2} & f_{X_3X_3} \end{bmatrix}$$

F_{ij} = determinant of the ij^{th} cofactor of the above bordered Hessian

Both preceding indices measure the effects of movements along an indifference surface. In addition to these, some measures of the positioning of one surface (at a given level of utility) relative to another is essential in the measurement of the effects of increased income upon the demand structure. The Hicks-Allen coefficients of income variation perform just such a function. Equations (23) through (25) define them for each of the three goods of interest:

$$\rho_{X_1} = \frac{X_2 \cdot X_3}{R_{X_1}{}^{X_2} \cdot R_{X_1}{}^{X_3}} \begin{vmatrix} \dfrac{\partial}{\partial X_2}(R_{X_1}{}^{X_2}) & \dfrac{\partial}{\partial X_3}(R_{X_1}{}^{X_2}) \\ \dfrac{\partial}{\partial X_2}(R_{X_1}{}^{X_3}) & \dfrac{\partial}{\partial X_3}(R_{X_1}{}^{X_3}) \end{vmatrix} \tag{23}$$

$$\rho_{X_2} = -\frac{X_1 \cdot X_3}{R_{X_1}{}^{X_2} \cdot R_{X_1}{}^{X_3}} \begin{vmatrix} \dfrac{\partial}{\partial X_1}(R_{X_1}{}^{X_2}) & \dfrac{\partial}{\partial X_3}(R_{X_1}{}^{X_2}) \\ \dfrac{\partial}{\partial X_1}(R_{X_1}{}^{X_3}) & \dfrac{\partial}{\partial X_3}(R_{X_1}{}^{X_3}) \end{vmatrix} \tag{24}$$

$$\rho_{X_3} = \frac{X_1 \cdot X_2}{R_{X_1}{}^{X_2} \cdot R_{X_1}{}^{X_3}} \begin{vmatrix} \dfrac{\partial}{\partial X_1}(R_{X_1}{}^{X_2}) & \dfrac{\partial}{\partial X_2}(R_{X_1}{}^{X_2}) \\ \dfrac{\partial}{\partial X_1}(R_{X_1}{}^{X_3}) & \dfrac{\partial}{\partial X_2}(R_{X_1}{}^{X_3}) \end{vmatrix} \tag{25}$$

These measures may be somewhat easier to interpret by referring to a diagram. Figure 13 maps the utility surface for X_1, X_2, and X_3 at utility level one with $U_1U_1'U_1''$. If we are interested in measuring the curvature at a point on the surface, such as B, then we are interested in σ^*. Alternatively, if we hold the level of X_2 constant (at zero in our illustration), the partial elasticity of substitution, σ_{13} in this case, is the relevant measure. In the diagram, it is given by the rate of change in the slope of OA relative to the slope of a tangent to A. Finally, to interpret the coefficients of income variation, assume $U_1'\ U_1''$ is the cross-section of the utility surface at level one when X_2 is held at zero, and that $U_2'\ U_2''$ is the cross of the utility surface at level two when X_2 remains at zero. The coefficients of income variation hold one commodity constant and examine the effects of changes in the other two. For example, movement from E to R in figure 13 alters X_1 by HG and X_2 by FS to a level given by OT while X_3 is held constant at OF. This can be broken down into two components. First, the variation in X_1 results in a movement from E to a higher level of utility $U_2''\ U_2'$ in the X_1X_3 plane at point D. Second, the variation in X_2 moves the level of utility to $U_3''\ U_3'$ in another X_1X_3

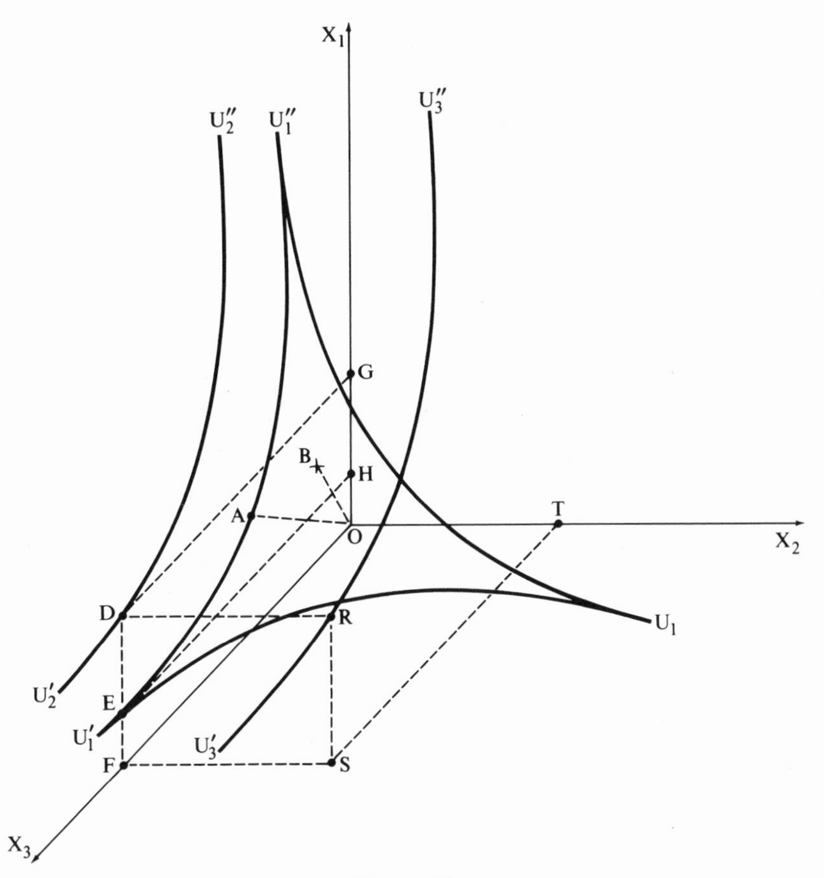

FIGURE 13.

plane with X_2 at OT and the point designated by R. These two movements are sufficient to describe the spacing of the indifference surfaces for X_3 and are given in the Hicks-Allen coefficient of income variation. In the same manner ρ_{X_1} and ρ_{X_2} provide information about the spacing of the indifference surfaces in the X_1 and X_2 directions, respectively.

Finally, the elasticity of substitution between any one good and the remaining pair (i.e., for X_1, $\sigma^*/_{X_2X_3}\sigma_{X_2X_3}$) can be defined in terms of the partial elasticities of substitution on the proportion of the budget spent on each of the three commodities. The equations necessary for such a definition are as follows:

$$(1 - K_{X_1}) \frac{\sigma^*}{_{X_2X_3}\sigma_{X_2X_3}} - K_{X_2}\sigma_{X_1X_2} - K_{X_3}\sigma_{X_1X_3} = 0 \qquad (26)$$

$$(1 - K_{X_2}) \frac{\sigma^*}{{}_{X_1X_3}\sigma_{X_1X_3}} - K_{X_1}\sigma_{X_1X_2} - K_{X_3}\sigma_{X_2X_3} = 0 \qquad (27)$$

$$(1 - K_{X_3}) \frac{\sigma^*}{{}_{X_1X_2}\sigma_{X_1X_2}} - K_{X_1}\sigma_{X_1X_2} - K_{X_2}\sigma_{X_2X_3} = 0 \qquad (28)$$

where: $K_{X_1} = \dfrac{P_{X_1} \cdot X_1}{y}$

$K_{X_2} = \dfrac{P_{X_2} \cdot X_2}{y}$

$K_{X_3} = \dfrac{P_{X_3} \cdot X_3}{y}$

Several definitions may now be stated and then the equations linking income, price, and cross elasticities of demand (e.g., the elasticity analogs of the Slutsky equations).

$$K_{X_1} + K_{X_2} + K_{X_3} = 1 \qquad (29)$$

$$K_{X_1}\rho_{X_1} + K_{X_2}\rho_{X_2} + K_{X_3}\rho_{X_3} = 1/\sigma^* \qquad (30)$$

Equations (31) through (33) define the income elasticities of demand:

$$E_y(X_1) = \sigma^* \cdot \rho_{X_1} \qquad (31)$$

$$E_y(X_2) = \sigma^* \cdot \rho_{X_2} \qquad (32)$$

$$E_y(X_3) = \sigma^* \cdot \rho_{X_3} \qquad (33)$$

where: $E_y(X_i) = \dfrac{\partial X_i}{\partial y} \cdot \dfrac{y}{X_i}$; $i = 1, 2, 3$

Given this equipment, and assuming that the individual maximizes his utility subject to a budget constraint and within the framework of traditional neoclassical analysis, Hicks and Allen derive the following relationships:

$$E_{P_{X1}}(X_1) = K_{X_1}E_y(X_1) + (1 - K_{X_1}) \frac{\sigma^*}{{}_{X_2X_3}\sigma_{X_2X_3}} \qquad (34)$$

$$E_{P_{X1}}(X_2) = K_{X_1}E_y(X_2) - K_{X_1}\sigma_{X_1X_2} \qquad (35)$$

$$E_{P_{X1}}(X_3) = K_{X_1}E_y(X_3) - K_{X_1}\sigma_{X_1X_3} \qquad (36)$$

$$E_{P_{X2}}(X_2) = K_{X_2}E_y(X_2) + (1 - K_{X_2}) \frac{\sigma^*}{{}_{X_1X_3}\sigma_{X_1X_3}} \qquad (37)$$

$$E_{P_{X2}}(X_1) = K_{X_2}E_y(X_1) - K_{X_2}\sigma_{X_1X_2} \qquad (38)$$

$$E_{P_{X_2}}(X_3) = K_{X_2}E_y(X_3) - K_{X_2}\sigma_{X_2X_3} \tag{39}$$

$$E_{P_{X_3}}(X_3) = K_{X_3}E_y(X_3) + (1 - K_{X_3})\frac{\sigma^*}{_{X_1X_2}\sigma_{X_1X_2}} \tag{40}$$

$$E_{P_{X_3}}(X_1) = K_{X_3}E_y(X_1) - K_{X_3}\sigma_{X_1X_3} \tag{41}$$

$$E_{P_{X_3}}(X_2) = K_{X_3}E_y(X_2) - K_{X_3}\sigma_{X_2X_3} \tag{42}$$

where: $E_{P_{X_i}}(X_j) = -\dfrac{\partial X_j}{\partial P_{X_i}} \cdot \dfrac{P_{Xi}}{X_j}; \quad \begin{matrix} j = 1, 2, 3 \\ i = 1, 2, 3 \end{matrix}$

All these equations permit one to characterize the demand structure implied by any utility function. Consequently, by examining a variety of specifications for utility, we can select a utility function that conforms as nearly as possible to the pattern of demand observed in the "real world" for the goods and services of immediate interest.

In order to illustrate the derivation of the properties for one utility function, consider a Cobb-Douglas specification, as in equation (43):[4]

$$U = AX_1^{\alpha_1}X_2^{\alpha_2}X_3^{\alpha_3} \tag{43}$$

To evaluate the pattern of demand implied, we must systematically evaluate each of the Hicks-Allen indices of preferences. First, we construct the marginal rates of substitution:

$$\frac{\partial U}{\partial X_1} = \alpha_1 AX_1^{(\alpha_1-1)}X_2^{\alpha_2}X_3^{\alpha_3} \tag{44a}$$

$$\frac{\partial U}{\partial X_2} = \alpha_2 AX_1^{\alpha_1}X_2^{(\alpha_2-1)}X_3^{\alpha_3} \tag{44b}$$

$$\frac{\partial U}{\partial X_3} = \alpha_3 AX_1^{\alpha_1}X_2^{\alpha_2}X_3^{(\alpha_3-1)} \tag{44c}$$

$$R_{X_1}{}^{X_2} = -\frac{dX_1}{dX_2} = \frac{\dfrac{\partial U}{\partial X_2}}{\dfrac{\partial U}{\partial X_1}} = \frac{\alpha_2}{\alpha_1}\frac{X_1}{X_2} \tag{45}$$

$$R_{X_1}{}^{X_3} = -\frac{dX_1}{dX_3} = \frac{\dfrac{\partial U}{\partial X_3}}{\dfrac{\partial U}{\partial X_1}} = \frac{\alpha_3}{\alpha_1}\frac{X_1}{X_3} \tag{46}$$

[4] There has been increasing concern with the implications of a given utility function for a pattern of demand, and vice versa. A good recent example is Sato (1972).

Our index of curvature, σ^*, is given as follows:

$$\sigma^* = \frac{\dfrac{\alpha_2}{\alpha_1}\dfrac{X_1}{X_2}\dfrac{\alpha_3}{\alpha_1}\dfrac{X_1}{X_3}}{X_1 \cdot X_2 \cdot X_3} \cdot \frac{X_1 + \dfrac{\alpha_2}{\alpha_1}X_1 + \dfrac{\alpha_3}{\alpha_1}X_1}{G} \tag{47}$$

and

$$G = \det \begin{bmatrix} 1 & \dfrac{\alpha_2}{\alpha_1}\dfrac{X_1}{X_2} & \dfrac{\alpha_3}{\alpha_1}\dfrac{X_1}{X_3} \\ \dfrac{\alpha_2}{\alpha_1}\dfrac{1}{X_2} & -\dfrac{\alpha_2}{\alpha_1}\dfrac{X_1}{X_2^2} & 0 \\ \dfrac{\alpha_3}{\alpha_1}\dfrac{1}{X_3} & 0 & -\dfrac{\alpha_3}{\alpha_1}\dfrac{X_1}{X_3^2} \end{bmatrix}$$

$$G = \frac{\alpha_2\alpha_3}{\alpha_1^3}(\alpha_1 + \alpha_2 + \alpha_3)\frac{X_1^2}{X_2^2X_3^2}$$

Simplifying (47), we have:

$$\sigma^* = \frac{\alpha_2\alpha_3}{\alpha_1^3}(\alpha_1 + \alpha_2 + \alpha_3)\frac{X_1^2}{X_2^2X_3^2} \cdot \frac{1}{G} = 1 \tag{48}$$

The Cobb-Douglas function is rather unusual in form; not only is the overall index of curvature, σ^*, equal to unity, but partial elasticities are also unity. Rather than provide the rather tedious manipulation in all cases, we will demonstrate this proposition in one case.

From equation (22), we define F for the Cobb-Douglas case as the determinant of the matrix given in (49):

$$\begin{bmatrix} 0 & \alpha_1AX_1^{\alpha_1-1}X_2^{\alpha_2}X_3^{\alpha_3} & \alpha_2AX_1^{\alpha_1}X_2^{\alpha_2-1}X_3^{\alpha_3} & \alpha_3AX_1^{\alpha_1}X_2^{\alpha_2}X_3^{\alpha_3-1} \\ \alpha_1AX_1^{\alpha_1-1}X_2^{\alpha_2}X_3^{\alpha_3} & \alpha_1(\alpha_1-1)AX_1^{\alpha_1-2}X_2^{\alpha_2}X_3^{\alpha_3} & \alpha_1\alpha_2AX_1^{\alpha_1-1}X_2^{\alpha_2-1}X_3^{\alpha_3} & \alpha_1\alpha_3AX_1^{\alpha_1-1}X_2^{\alpha_2}X_3^{\alpha_3-1} \\ \alpha_2AX_1^{\alpha_1}X_2^{\alpha_2-1}X_3^{\alpha_3} & \alpha_1\alpha_2AX_1^{\alpha_1-1}X_2^{\alpha_2-1}X_3^{\alpha_3} & \alpha_2(\alpha_2-1)AX_1^{\alpha_1}X_2^{\alpha_2-2}X_3^{\alpha_3} & \alpha_2\alpha_3AX_1^{\alpha_1}X_2^{\alpha_2-1}X_3^{\alpha_3-1} \\ \alpha_3AX_1^{\alpha_1}X_2^{\alpha_2}X_3^{\alpha_3-1} & \alpha_1\alpha_3AX_1^{\alpha_1-1}X_2^{\alpha_2}X_3^{\alpha_3-1} & \alpha_2\alpha_3AX_1^{\alpha_1}X_2^{\alpha_2-1}X_3^{\alpha_3-1} & \alpha_3(\alpha_3-1)AX_1^{\alpha_1}X_2^{\alpha_2}X_3^{\alpha_3-2} \end{bmatrix} \tag{49}$$

If we examine the value of F and the determinant of the (1, 2) cofactor of the matrix above, their ratio (F_{12}/F) is equal to the following:

$$\frac{F_{12}}{F} = (X_1X_2)/(X_1\alpha_1AX_1^{\alpha_1-1}X_2^{\alpha_2}X_3^{\alpha_3} + X_2\alpha_2AX_1^{\alpha_1}X_2^{\alpha_2-1}X_3^{\alpha_3} + X_3\alpha_3AX_1^{\alpha_1}X_2^{\alpha_2}X_3^{\alpha_3-1}) \tag{50}$$

Accordingly, $\sigma_{12} = 1$, and it can be demonstrated that all partial elasticities are also unity. The last components necessary to evaluate the demand structure implied by a Cobb-Douglas utility function are the coefficients of income variation, which may be derived from their definitions in equations (23) through (25), as in equations (51) through (53):

$$\rho_{X_1} = \frac{X_2 X_3}{\left(\frac{\alpha_2}{\alpha_1}\frac{X_1}{X_2}\right)\left(\frac{\alpha_3}{\alpha_1}\frac{X_1}{X_3}\right)} \cdot \begin{vmatrix} -\frac{\alpha_2}{\alpha_1}\frac{X_1}{X_2^2} & 0 \\ 0 & \frac{\alpha_3}{\alpha_1}\frac{X_1}{X_3^2} \end{vmatrix} = 1 \tag{51}$$

$$\rho_{X_2} = \frac{-X_1 X_3}{\left(\frac{\alpha_2}{\alpha_1}\frac{X_1}{X_2}\right)\left(\frac{\alpha_3}{\alpha_1}\frac{X_1}{X_3}\right)} \cdot \begin{vmatrix} \frac{\alpha_2}{\alpha_1}\frac{1}{X_2} & 0 \\ \frac{\alpha_3}{\alpha_1}\frac{1}{X_3} & -\frac{\alpha_3}{\alpha_1}\frac{X_1}{X_3^2} \end{vmatrix} = 1 \tag{52}$$

$$\rho_{X_3} = \frac{X_1 X_2}{\left(\frac{\alpha_2}{\alpha_1}\frac{X_1}{X_2}\right)\left(\frac{\alpha_3}{\alpha_1}\frac{X_1}{X_3}\right)} \cdot \begin{vmatrix} \frac{\alpha_2}{\alpha_1}\frac{1}{X_2} & -\frac{\alpha_2}{\alpha_1}\frac{X_1}{X_2^2} \\ \frac{\alpha_3}{\alpha_1}\frac{1}{X_3} & 0 \end{vmatrix} = 1 \tag{53}$$

We now have sufficient information to describe completely the demand structure implied by a Cobb-Douglas utility specification. First, the income elasticities are everywhere unity by equations (31) through (33) and the results in equations (48) and (51) through (53). Second, the price elasticities of demand are also unity, though this is not immediately obvious. Therefore, consider an example $E_{P_{X_1}}(X_1)$:

$$(1 - K_{X_1}) \frac{\sigma^*}{{}_{X_2X_3}\sigma_{X_2X_3}} = K_{X_2} + K_{X_3} \tag{54}$$

Substituting into equation (34) the value for the income elasticity and that of $\sigma^*/{}_{X_2X_3}\sigma_{X_2X_3}$, we have:

$$E_{P_{X_1}}(X_1) = K_{X_1}(1) + K_{X_2} + K_{X_3} = 1 \tag{55}$$

The same relationship holds for the other price elasticities. The cross elasticities are zero, since $E_y(X_i) = \sigma_{X_iX_j}$ for all i and j.

The demand structure implied by a Cobb-Douglas utility specification is exceptionally rigid. Unitary income and price elasticities with no cross-effects are hardly characteristic of any set of three commodities in the "real world." Moreover, our brief remarks on the nature of the demand for amenity services indicate that this function is inappropriate if such services are to be one of the three goods.

Several alternative utility specifications thus have been considered.

Three additional utility functions are given in equations (56) through (58):

$$U = A(a_1X_1^{-\beta} + a_2X_2^{-\beta} + a_3X_3^{-\beta})^{-1/\beta} \tag{56}$$

$$U = A(a_1X_1^{-\beta} + a_2X_2^{-\beta})^{-p/\beta} \cdot a_3X_3^{q},\ p + q = 1 \tag{57}$$

$$U = A(a_1X_1^{-\beta_1} + a_2X_2^{-\beta_2} + a_3X_3^{-\beta_3})^{-1/\beta} \tag{58}$$

The first of these is the constant elasticity of substitution (*CES*) utility function originally introduced by Arrow and others (1961) in their study of international production patterns. The second is a constant "in pairs" utility function (hereafter designated *CES′*) introduced by Uzawa (1962) and discussed more recently for production relationships by Ferguson (1969). The last is Mukerji's (1963) constant ratio of partial elasticity of substitution function, which reduces to Houthakker's (1960) additive preferences function when $\beta = -1$.

The *CES* function has partial elasticities of substitution which equal $1/(1 + \beta)$. This is also the value of the total elasticity, σ^*. In contrast, the *CES′* function has differing partial elasticities of substitution, in that $\sigma_{12} = 1/(1 + \beta)$ and $\sigma_{13} = \sigma_{23} = 1$. Moreover, the total elasticity, σ^*, in this case equals σ_{12}. Finally, equations (59) and (60) provide the partial elasticities of substitution and total elasticity with the Mukerji function:

$$\frac{\sigma_{12}}{\sigma_{13}} = \frac{1 + \beta_3}{1 + \beta_2};\ \frac{\sigma_{23}}{\sigma_{12}} = \frac{1 + \beta_1}{1 + \beta_3} \tag{59}$$

$$\sigma^* = \left[1 + \frac{X_2}{X_1}(R_{X_1}{}^{X_2}) + \frac{X_3}{X_1}(R_{X_1}{}^{X_2})\right] \Big/ \Big[(1 + \beta_3)(1 + \beta_2) + (1 + \beta_3)(1 + \beta_1)\frac{X_2}{X_1}(R_{X_1}{}^{X_2}) + (1 + \beta_2)(1 + \beta_1)\frac{X_3}{X_1}(R_{X_1}{}^{X_3})\Big] \tag{60}$$

Given this information on each of the three utility functions of interest and the Hicks-Allen equations previously stated, we can elaborate the demand structure of each of these functions. Tables 1, 2, and 3 provide such a description, giving the results obtained by working through the Hicks-Allen equations for each case.

Table 1. Structure of Demand: *CES* Utility Function

Commodity	$E_{Pi}(X_1)$	$E_{Pi}(X_2)$	$E_{Pi}(X_3)$	$E_y(X_i)$
X_1	$K_1(1 - \sigma) + \sigma$	$K_1(1 - \sigma)$	$K_1(1 - \sigma)$	1
X_2	$K_2(1 - \sigma)$	$K_2(1 - \sigma) + \sigma$	$K_2(1 - \sigma)$	1
X_3	$K_3(1 - \sigma)$	$K_3(1 - \sigma)$	$K_3(1 - \sigma) + \sigma$	1

NOTE: K_i = the proportion of the budget spent on X_i

$$\sigma = \frac{1}{1 + \beta}$$

Table 2. Structure of Demand: CES' Utility Function

Commodity	$E_{P_i}(X_1)$	$E_{P_i}(X_2)$	$E_{P_i}(X_3)$	$E_y(X_i)$
X_1	$K_1 + K_3 + K_2\sigma_{12}$	$K_1(1 - \sigma_{12})$	0	1
X_2	$K_2(1 - \sigma_{12})$	$K_2 + K_3 + K_1\sigma_{12}$	0	1
X_3	0	0	1	1

NOTE: K_i = proportion of the budget spent on X_i

$$\sigma_{12} = \frac{1}{1 + \beta}$$

The results in the first two tables indicate that both the CES and CES' utility specifications require fairly rigid demand patterns to be obeyed by the goods and services designated X_1, X_2, and X_3. The unitary income elasticities are to be expected with any homogeneous utility specification.

When amenity services are among the goods and services desired by the community, however, this assumption is untenable. For this reason alone, we can eliminate any homogeneous specification of utility from serious consideration. A nonhomogeneous format eliminates some of these disadvantages. However, there is a wide range of possible choices. Our results relate to the one that seems the most appropriate, a generalization of Houthakker's (1960) specification of additive preferences. Moreover, Sato's (1972) results indicate that it corresponds approximately to the utility function implied by many double-log empirical analyses of the demand structure for certain commodities. Finally, as table 3 indicates, this specification introduces a most flexible demand structure, capable of accommodating a fairly broad array of alternative goods and services.

Income elasticities must remain constant in ratio in the same proportion as the partial elasticities of substitution. In addition, the absolute magnitude of price and cross elasticities responds to the level of consumption. Moreover, it is possible to encompass the other specifications' results within this framework.

SUMMARY

There are two sides of a general equilibrium model. One describes the nature of supply and for our model was presented in chapter 3. The other describes demand. In this chapter we have outlined the empirical and theoretical considerations in the choice of a means of specifying a community's demand structure.

Table 3. Structure of Demand: Mukerji's Utility Function

Commodity	$E_{Pi}(X_1)$	$E_{Pi}(X_2)$	$E_{Pi}(X_3)$	$E_y(X_i)$
X_1	$K_1(1+\beta_3)(1+\beta_2)\sigma^* + \sigma_{13}\left(K_2\frac{1+\beta_3}{1+\beta_2}+K_3\right)$	$K_1(1+\beta_1)(1+\beta_3)\sigma^* - K_1\sigma_{13}\left(\frac{1+\beta_3}{1+\beta_2}\right)$	$K_1(1+\beta_1)(1+\beta_2)\sigma^* - K_1\sigma_{13}$	$(1+\beta_2)(1+\beta_3)\sigma^*$
X_2	$K_2(1+\beta_2)(1+\beta_3)\sigma^* - K_2\left(\frac{1+\beta_3}{1+\beta_2}\right)\sigma_{13}$	$K_2(1+\beta_1)(1+\beta_3)\sigma^* + \sigma_{13}\left(K_1\frac{1+\beta_3}{1+\beta_2}+K_3\frac{1+\beta_1}{1+\beta_2}\right)$	$K_2(1+\beta_1)(1+\beta_2)\sigma^* - K_2\left(\frac{1+\beta_1}{1+\beta_2}\right)\sigma_{13}$	$(1+\beta_1)(1+\beta_3)\sigma^*$
X_3	$K_3(1+\beta_2)(1+\beta_3)\sigma^* - K_3\sigma_1$	$K_3(1+\beta_1)(1+\beta_3)\sigma^* - K_3\left(\frac{1+\beta_1}{1+\beta_2}\right)\sigma_{13}$	$K_3(1+\beta_1)(1+\beta_2)\sigma^* + \sigma_{13}\left(K_2\frac{1+\beta_1}{1+\beta_2}+K_1\right)$	$(1+\beta_2)(1+\beta_1)\sigma^*$

NOTE: σ^* is defined in equation (57)
K_i = proportion of the budget spent on commodity X_i

(1) One of the most important types of amenity services to be derived from environmental resources in their preserved status is outdoor recreation. The empirical studies of the nature of the demand for such services indicate that such services tend to become luxury goods. Moreover, many of these services do not have adequate substitutes.

(2) The approach taken in our model for the specification of community preferences is to postulate a community utility function. This function can be rationalized in one of three ways: (a) the restrictive model used for *CIC* (community indifference curves), wherein all individuals are identical; (b) as a function that provides an alternative mode of specifying a set of demand relationships and nothing more; or (c) under the guise of being associated with some "representative" individual in the community.

(3) There are two means of examining the effect of variations in income or in a commodity or service's price upon the nature of demand. The first is through the use of Slutsky equations; the second makes use of the Hicks-Allen relationships. Both present equivalent pictures.

(4) For the present purposes, Hicks-Allen equations offer a more convenient mode for analyzing the effects of price and income variations, since they are expressed in terms of elasticities on which there is typically more prior information.

(5) Four utility specifications have been examined: Cobb-Douglas; *CES*; a constant partial elasticity of substitution "in pairs" function; and Mukerji's specification. The last provides the most flexible demand structure.

REFERENCES

Archibald, G. C. 1965. "The Qualitative Content of Maximizing Models." *Journal of Political Economy* 73: 27–36.

Arrow, K. J., et al. 1961. "Capital Labor Substitution and Economic Efficiency." *Review of Economics and Statistics* 43: 225–250.

Baumol, W. J. 1949–50. "The Community Indifference Map." *Review of Economic Studies* 17: 189–197.

Burt, O. R., and D. Brewer. 1971. "Estimation of Net Social Benefits from Outdoor Recreation." *Econometrica* 39: 813–827.

Chipman, J. S., et al., eds. 1971. *Preferences, Utility, and Demand.* New York: Harcourt Brace Jovanovich.

Cicchetti, C. J. 1972. "A Multivariate Statistical Analysis of Wilderness Users in the United States." In *Natural Environments: Studies in Theoretical and Applied Analysis*, ed. J. V. Krutilla. Baltimore: The Johns Hopkins University Press for Resources for the Future, Inc.

Cicchetti, C. J., A. C. Fisher, and V. K. Smith. 1973. "Economic Models and Planning Outdoor Recreation." *Operations Research* 21: 1104–13.

Cicchetti, C. J., J. J. Seneca, and P. Davidson. 1969. *The Demand and Supply of Outdoor Recreation*. New Brunswick, N.J.: Bureau of Economic Research.

Cicchetti, C. J., et al. 1972. "Recreation Benefit Estimation and Forecasting: The Implications of the Identification Problem." *Water Resources Research* 8: 840–850.

Clawson, M. 1959. "Methods of Measuring Demand for and Value of Outdoor Recreation." RFF Reprint 10.

Davidson, P., F. G. Adams, and J. Seneca. 1966. "The Social Value of Water Recreational Facilities Resulting from an Improvement in Water Quality: The Delaware Estuary." In *Water Research*, ed. A. V. Kneese and S. C. Smith. Baltimore: The Johns Hopkins Press.

Ferguson, C. E. 1969. *The Neoclassical Theory of Production and Distribution*. Cambridge: Cambridge University Press.

Fisher, F. M. 1972. "Gross Substitutes and the Utility Function." *Journal of Economic Theory* 4: 82–87.

Goldberger, A. S. 1967. "Functional Form and Utility: A Review of Consumer Demand Theory." Social Systems Research Institute, University of Wisconsin. Unpublished.

Gorman. W. M. 1953. "Community Preference Fields." *Econometrica* 21: 63–80.

Heller, H. R. 1968. *International Trade: Theory and Empirical Evidence*. Englewood Cliffs, N.J.: Prentice-Hall, Inc.

Henderson, J. M., and R. E. Quandt. 1958. *Microeconomic Theory: A Mathematical Approach*. New York: McGraw-Hill

Hicks, J. R., and R. G. D. Allen. 1934. "A Reconsideration of the Theory of Value—Parts I and II." *Economica* 11: 52–77, 196–219.

Houthakker, H. S. 1960. "Additive Preferences." *Econometrica* 28: 244–257.

Katzner, D. W. 1970. *Static Demand Theory*. New York: Macmillan Co.

Lancaster, K. 1968. *Mathematical Economics*. New York: Macmillan Co.

Mukerji, V. 1963. "A Generalized SMAC Function with Constant Ratios of Elasticity of Substitution." *Review of Economic Studies* 30: 233–236.

Parks, R. W. 1969. "Systems of Demand Equations: An Empirical Comparison of Alternative Functional Forms." *Econometrica* 37: 629–650.

Pearse, P. H. 1968. "A New Approach to the Evolution of Non-Priced Recreational Resources." *Land Economics* 44: 87–100.

Powell, A. A. 1971. *Informative Estimation of Economic Models: The Empirical Analytics of Demand and Supply*. Clayton, Australia: Monash University.

Quirk, J., and R. Saposnik. 1968. *Introduction to General Equilibrium Theory and Welfare Economics*. New York: McGraw-Hill.

Samuelson, P. A. 1956. "Social Indifference Curves." *Quarterly Journal of Economics* 70: 1–22.

Sato, K. 1972. "Additive Utility Function with Double-Log Consumer Demand Functions." *Journal of Political Economy* 80: 103–124.

Seneca, J. J., P. Davidson, and F. G. Adams. 1968. "An Analysis of Recreational Use of the TVA Lakes." *Land Economics* 44: 529–534.

Slutsky, E. E. 1915. "On the Theory of the Budget of the Consumer." *Giornale degli Economisti*, Vol. 51 (July 1915). Reprinted in *Readings in Price Theory*. Homewood, Ill.: Irwin, 1952.

Stankey, G. 1972. "A Strategy for the Definition and Management of Wilderness Quality." In *Natural Environments: Studies in Theoretical and Applied Analysis*, ed. J. V. Krutilla. Baltimore: The Johns Hopkins University Press for Resources for the Future, Inc.

Tolley, G. S. 1965. "Recreation Benefits from Water Pollution Control." *Water Resources Research* (Second Quarter).

Uzawa, H. 1962. "Production Functions with Constant Elasticities of Substitution." *Review of Economic Studies* 29: 291–299.

chapter five Technical Change and Relative Prices

As chapter 1 indicated, natural endowments are the given and fixed source of all goods and services on which the community depends. Technical change, as Barnett and Morse (1963) have argued, has allowed for an ever-expanding supply of most natural resource commodities at constant or falling supply prices. While the standard of living, measured in terms of per capita consumption, has been increasing, some dimensions of the quality of life have deteriorated. Our explanation for this asymmetry is based on the differing effects of technical change on the uses of the services of natural endowments.

These services may function as factor inputs contributing to the production of final consumption goods or, alternatively, they may, with little transformation, enter into the utility function of final consumers as amenity services. Technical change tends to favor the first allocation, leaving the second unaffected. Accordingly, one suspects their value in the latter use will, over time, increase. The intertemporal effects of using such services to produce fabricated goods are not included in the analysis at this point.

The first section of this chapter discusses the primary assumptions and the strategy of the model. Section two reviews the model with two goods and a homogeneous utility specification. The third section considers the model with a nonhomogeneous function. It is followed by consideration of a three-commodity version of the model and a brief summary of the results of these models.

ASSUMPTIONS AND STRATEGY

General equilibrium models are typically easy to formulate in concept, but operationally difficult to work with. On balance, however, they are the most useful for our purposes. The building blocks for these models have been defined in chapters 3 and 4; we will work with "summary functions" for the supply and demand sides of the markets in the model. Rather than consider each individual in the community, we will summarize the whole of the society's demand structure, using a utility function as described in chapter 4. A similar capsulization of supply into production transformation functions was described in chapter 3. Our concern here is with the movement in relative prices as we alter the nature of our assumed demand structure and the specified technical change.

The simplest means for analyzing the asymmetric implications of technology is to consider their effects in the context of a two-commodity model. Krutilla (1967) has provided the framework and the key components to be considered in such a conceptualization. Assume one of these two goods to be the amenity services of unspoiled natural environments. An example of such an environmental resource is a wilderness area with unique attributes, such as Hells Canyon. When the area is used for certain recreational pursuits or for scientific purposes, its services can be neither expanded nor reproduced. Increased productivity as a result of technological change is not possible with such services. By way of contrast, we shall assume the other good in the two-sector model to be a fabricated or manufactured good. Technical change does, in this case, act to increase the available supply over time. Consequently, a dichotomy has been established within the supply side of the model.

In order to model the impact of this asymmetry in the effect of technical change, several restrictive assumptions will be made in addition to those stated in chapters 3 and 4, where the belief that the world is "neoclassical" is implicit in the analysis. The supply of factor inputs will be assumed constant, and all inputs to be fully employed. Technical change

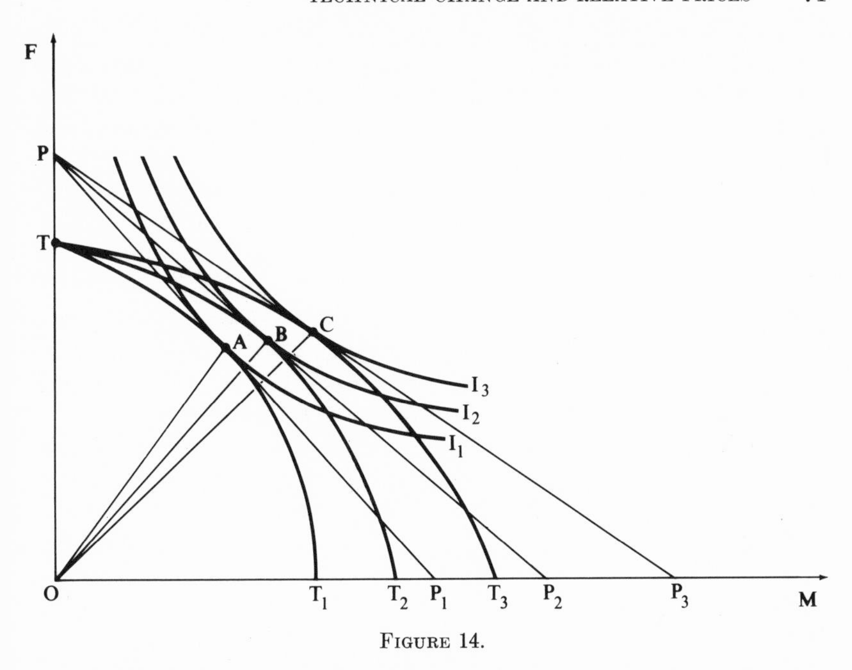

FIGURE 14.

is disembodied from the factor inputs with its magnitude and direction as given. Finally, we shall assume that consumption and production externalities of the traditional character are absent.[1] (Consideration of the specific intertemporal externality of concern to us will be deferred until chapter 6. The irreversibility aspects of certain development decisions thus will not be confronted in this chapter.)

Before we describe the objectives of our model more specifically, a word on notation is in order. The notation thus far has been fairly general, with no specific meaning consistently assigned to a symbol. For this chapter and the next, we shall designate manufactured goods by the symbol M and the amenity services of natural endowments by F. When a third good is introduced, it will be designated X, without specific meaning. (We will wish to examine alternative scenarios corresponding to different meanings for X.)

Figure 14 describes our model in two dimensions. The curves I_1, I_2, and I_3 are excerpts from the community's utility map describing its preferences for F and M. The curves TT_1, TT_2, and TT_3 are three production transformation curves corresponding to different time periods and resultant technical change. The shift from TT_1 to TT_2 illustrates

[1] For a brief discussion of the effect of the common property nature of environmental services upon the production possibility curve, see chapter 3.

the pattern of technical change in which the production of fabricated goods is progressively favored over time, stretching the terminus of the curve on the M axis. The comparative static equilibrium points in the face of such technical change are given by A, B, and C. The reasoning underlying this conclusion is straightforward. These points are combinations of M and F where the marginal rate of substitution in consumption and the marginal rate of transformation from production are equal. Consequently, all resources are fully employed, and there are no gains to be realized by reallocations between the production of M and F, or, on the demand side, by choosing an alternative combination.

Corresponding to each of these equilibrium points is a commodity mix and price ratio. The price ratio can be determined by examining the slope of tangents to the points A, B, and C. These tangents are given in figure 14 by PP_1, PP_2, and PP_3, respectively. Our primary interest is in how the slope changes with the technical change and a given structure of demand. For example, in the figure the price of M relative to F clearly is declining as we move from PP_1 to PP_2 to PP_3. The rate of this decline relative to the rate of technical change is not clear, however. Furthermore, we cannot ascertain the effect of different demand structures with a preassigned technical change upon this rate.

The commodity mix (ratio of F to M) consumed at each equilibrium point is also of interest. This ratio can be discerned by examining the slopes of rays from the origin to the equilibrium point in question, i.e., OA, OB, and OC, respectively. We find that the ratio of F to M is progressively declining, indicating that the community is increasing the quantity of M relative to F it consumes.

TWO-GOOD MODEL WITH A HOMOGENEOUS UTILITY SPECIFICATION

Graphical analysis of the effects of technical change upon relative prices, while instructive, has some severe limitations. It does not, for example, permit the rate of relative price change to be linked to the rate of technical change. Consequently we shall specify a form for the utility mapping, recognizing its assumed demand structure, and one for the transformation curve. Equation (1) specifies a constant elasticity of substitution community indifference curve, similar in form to that used by Arrow and others (1961) in their study of international production patterns:

$$U = (a_1 M^{(1+B)/B} + a_2 F^{(1+B)/B})^{B/(1+B)} \tag{1}$$

where: U = level of community welfare
a_1, a_2 = non-neutral shift parameters
$B = -\sigma$, elasticity of substitution ($\sigma > 0$)

As noted in chapter 4, this format implies that the pattern of demand for manufactured goods (M) and amenity services (F) is quite restrictive. Income elasticity is constant at unity for all levels of consumption. Moreover, price and cross elasticities of demand are related to elasticity of substitution and to proportions of the budget spent upon each commodity (see table 1, p. 62, for a more explicit statement of these variables).

The second component of the model is the production possibility curve, which is specified to follow a constant elasticity of transformation (CET) form as defined by Powell and Gruen (1969). The implications of this specification, discussed in chapter 3, will not be elaborated here. In the two-commodity case, it is given in equation (2):

$$\alpha_1 M^{1-1/\tau} + \alpha_2 F^{1-1/\tau} = k \tag{2}$$

where: α_1, α_2 = non-neutral shift parameters
τ = elasticity of transformation ($\tau < 0$)
k = neutral shift parameter

Solving (1) for the marginal rate of substitution and (2) for the marginal rate of transformation results in equations (3) and (4), respectively:

$$MRS_{MF} = -\frac{dM}{dF} = \frac{a_2}{a_1}\left(\frac{M}{F}\right)^{1/\sigma} \tag{3}$$

$$MRT_{MF} = -\frac{dM}{dF} = \frac{\alpha_2}{\alpha_1}\left(\frac{M}{F}\right)^{1/\tau} \tag{4}$$

The comparative static equilibrium conditions call for equality of the marginal rate of substitution and marginal rate of transformation for a given commodity combination. In terms of figure 14, the CIC curve must be tangent to the relevant production possibility curve. This yields equations (5) and (6) for the commodity mix and relative prices, respectively:

$$\frac{M}{F} = \left(\frac{\alpha_2}{\alpha_1} \cdot \frac{a_1}{a_2}\right)^{\gamma} \tag{5}$$

$$\frac{P_F}{P_M} = \left(\frac{\alpha_2}{\alpha_1}\right)^{\theta}\left(\frac{a_1}{a_2}\right)^{\varphi} \tag{6}$$

where:

$$\gamma = \frac{\tau \cdot \sigma}{\tau - \sigma}; \gamma > 0$$

$$\theta = \frac{\tau}{\tau - \sigma}; \theta > 0$$

$$\varphi = \frac{\sigma}{\tau - \sigma}; \varphi < 0$$

Equations (5) and (6) describe, respectively, the slope of a ray such as OA and that of a tangent such as PP_1, in figure 14. Consequently, if we wish to determine the effect of technical change upon the commodity mix and price ratio in terms of the model's parameters, we must first provide an analytical description of change and then examine its effects upon (5) and (6).

As noted at the outset, technology tends to favor the production of fabricated goods (M) and not the amenity services (F). Accordingly, advances in technology will stretch the terminus of the production contour on the M axis while leaving that of F stationary. Assume for simplicity that this differential change proceeds at a constant rate, g; then the parameter ratio will be specified to follow an exponential form, as in (7):

$$\left(\frac{\alpha_2}{\alpha_1}\right)_t = \left(\frac{\alpha_2}{\alpha_1}\right)_0 e^{gt} \tag{7}$$

where:

$\left(\frac{\alpha_2}{\alpha_1}\right)_0$ = initial value of parameter ratio

g = rate of growth of technical change

t = time

e = base of natural logarithm system

It is clear that for an unchanging utility function both the commodity mix (M/F) and the price ration (P_F/P_M) will increase for positive g. An explicit expression for each may be derived by examining the rate of change of each as follows. Substituting from (7) into (5) and differentiating with respect to time, we have (8) as an expression for the rate of change in (M/F):

$$\frac{\frac{d(M/F)}{dt}}{M/F} = \frac{dM}{dt/M} - \frac{dF}{dt/F} = \gamma g \tag{8}$$

In terms of (5), this means that the equilibrium commodity ratio of M to F will increase (since $\gamma > 0$ and $g > 0$). In the limit as t is allowed to grow without bound, the consumption of M will also grow without bound. The case of F is somewhat more complex. Substituting from (5) into (2), we may solve for F in terms of the parameters of interest:[2]

$$F = \left[\frac{k}{1 + \left(\frac{a_1}{a_2}\right)^{\gamma(1-1/\tau)} \cdot \left(\frac{\alpha_2}{\alpha_1}\right)^{\tau(\sigma-1)/(\tau-\sigma)}} \right]^{\tau/(\tau-1)} \tag{9}$$

The limit of (9) can be determined by examining the exponent of (α_2/α_1), since this is the ratio which changes with t. The quantity of amenity services will approach zero only if the commodities M and F are substitutes (i.e., $\sigma > 1$). In the cases where they are independent (i.e., $\sigma = 1$) or complementary (i.e., $\sigma < 1$), F will approach a non-zero limit. Moreover, in the case of independence, the quantity of F will remain constant through time.

Clearly the key determinant of the commodity mixes chosen over time is the strength of the substitution effects. The nature of the *CES* utility function precludes examination of differential income effects over time. As a consequence, the results for this case may not be particularly valuable as guides for policy under real-world conditions. Nonetheless, they provide simple illustrations of the methodology.

The rate of change in relative prices may be determined by substituting for (α_2/α_1) in (6) and differentiating with respect to time. Equation (10) expresses the results:

$$\frac{\frac{d(P_F/P_M)}{dt}}{(P_F/P_M)} = \frac{\frac{dP_F}{dt}}{P_F} - \frac{\frac{dP_M}{dt}}{P_M} = \theta g \tag{10}$$

Relative prices will increase. However, θ will generally be less than or equal to one, so that the rate of relative price appreciation can be no greater than g. If M and F are perfect complements (i.e., $\sigma = 0$) in consumption, then the rate of relative price appreciation will equal the rate of biased technical change. Alternatively, if M and F are perfect substitutes (i.e., $\sigma \to \infty$), then fabricated goods and amenity services will exchange at a constant marginal rate of substitution for all possible combinations of consumption patterns for F and M, and there will be no relative price appreciation.

Supply considerations are equally important to the behavior of the commodity mix and relative price ratio. The elasticity of transformation, τ, measures the changes in the rate of transformation resulting from

[2] I am grateful to Clifford Russell and Karl-Göran Mäler for pointing out this approach and results.

changes in the desired commodity mix. It is, in other words, an index of curvature of the transformation frontier. Suppose for constant σ, the absolute value of τ is allowed to increase without bound. As a result there will be a constant marginal rate of transformation between M and F, and the rate of price appreciation will equal the rate of technnical change, g. This is the Baumol model described in chapter 2.

The limitations of this two-good homogeneous model are: (a) unitary income elasticities, (b) failure to consider substitutes for the amenity services, (c) rigid cross-elasticity effects, and (d) failure to consider irreversibilities resulting from past consumption decisions. In the next section an alteration in the form of the utility function will allow easing of the restrictions (a) and (c) found in the present model.

TWO-GOOD MODEL WITH NONHOMOGENEOUS UTILITY FUNCTION

Changing the community's utility function to a Mukerji (1963) constant ratio of partial elasticity of substitution function permits consideration of a more realistic demand structure. The income elasticities are not unity. Moreover, the price and cross elasticities are much less rigid in their relationship to the parameters of the model. That is, they can vary in numerical value with the levels of consumption of M and F (see table 3, p. 64). Equation (11) provides the function in its two-variable form:

$$U = (a_1 M^{-\beta_1} + a_2 F^{-\beta_2})^{-1/\beta} \tag{11}$$

Only when $\beta_1 = \beta_2$ is the function homogeneous. The marginal rate of substitution between M and F is as follows:

$$MRS_{MF} = \frac{a_2}{a_1} \cdot \frac{\beta_1}{\beta_2} \cdot \frac{M^{1+\beta_1}}{F^{1+\beta_2}} \tag{12}$$

With nonhomogeneous functions, solution of the system for the equilibrium commodity mix and price ratio provides results in terms of the parameters of the utility function, the parameters of the transformation function, and the level of consumption of one of the goods. Equating (12) and (4), we have (13) and (14):

$$M = \left(\frac{a_1}{a_2} \frac{\beta_1}{\beta_2} \frac{\alpha_2}{\alpha_1}\right)^{\gamma_1} F^{\gamma_2} \tag{13}$$

where:

$$\gamma_1 = \frac{1}{1 + \beta_1 - 1/\tau}\ ; \gamma_1 > 0$$

$$\gamma_2 = \frac{1 + \beta_2 - 1/\tau}{1 + \beta_1 - 1/\tau}\ ; \gamma_2 > 0$$

$$\frac{P_M}{P_F} = \left(\frac{\alpha_1}{\alpha_2}\right)^{(1+\beta_1)\gamma_1} \left(\frac{a_1\,\beta_1}{a_2\,\beta_2}\right)^{-\gamma_1/\tau} F^{(1-\gamma_2)/\tau} \tag{14}$$

The significance of the alteration in the assumed utility function is not apparent from a casual example of equations (13) and (14). However, assuming technical change to be proceeding under the conditions described in equation (7) and differentiating (14) with respect to time and deriving the rate of change in relative prices, we have equation (15). It should also be noted that this expression makes use of the relationship between the parameters of a Mukerji (1963) utility function and the demand structure derived in chapter 4.

$$\frac{\dfrac{d(P_M/P_F)}{dt}}{(P_M/P_F)} = \frac{E_y(F)}{\dfrac{\sigma}{\tau} - E_y(F)} \cdot g + \frac{E_y(M) - E_y(F)}{\sigma - E_y(F)\tau} \cdot \frac{\dot{F}}{F} \tag{15}$$

where:
$\dot{F} = dF/dt$
$E_y(F)$ = income elasticity of demand for F
$E_y(M)$ = income elasticity of demand for M

The first term on the right side of (15) is negative, since $\tau < 0$, $\sigma > 0$, $E_y(F) > 0$, and $g > 0$. Thus, if there is no increase in the equilibrium consumption of amenity services (i.e., $\dot{F} = 0$), the price ratio (P_M/P_F) will decline at a rate approaching g (the rate of technical change). The ratio of the elasticity of substitution to the elasticity of transformation gauges the effect of change. That is, σ measures the extent to which M and F are substitutes in terms of community preferences. τ assesses the facility with which the community may convert its resources between F and M at any given time.

The contribution of the second term to the rate of relative price change depends upon (1) the relative magnitude of the income elasticities and (2) the direction of change in the equilibrium consumption of F. Since amenity services are characteristically more income-elastic than fabricated goods and are likely to have positive growth in consumption, the second term will reinforce the effects of the first.

It is also interesting to compare these results with those derived under a homogeneous specification of the utility function. First, both income and substitution effects are important to relative price behavior in this case, whereas in the former model price change is related solely to "substitution"-like effects on the demand and supply effects. Second, the rate of growth of consumption of the environmental resources is important here, whereas in the homogeneous specification, relative price behavior was not affected by the rate at which the consumption of either good was expanding. The rationale for these separate effects stems from the nature of the Engel curves subsumed in the two specifications. In the homogeneous utility function, the Engel curves are all linear; with the nonhomogeneous specification, the Engel curves diverge from linearity. Consequently the level of consumption is important to the level of relative prices. However, this is just another way of saying that there are different income effects for the two goods.

THREE GOODS AND NONHOMOGENEOUS UTILITY FUNCTIONS

Generalization of the model to include a third commodity (designated X) that may serve as a substitute for, or complement to, either F or M adds both additional realism and complication to the model. Equations (16) and (17) define the generalized versions of the Mukerji utility function and the constant (partial) elasticity of transformation production possibility frontier (see chapters 3 and 4 for more detailed discussion).

$$U = (a_1 M^{-\beta_1} + a_2 F^{-\beta_2} + a_3 X^{-\beta_3})^{-1/\beta} \quad (16)$$

$$\alpha_1 M^{1-1/\tau} + \alpha_2 F^{1-1/\tau} + \alpha_3 X^{1-1/\tau} = k \quad (17)$$

Equating the marginal rates of substitution in consumption to the marginal rates of transformation, we derive expressions for the equilibrium commodity mix and price ratio for each pair of goods. Equations (18) and (19) present the commodity values, and (20) and (21) the price ratios.

$$M = \left(\frac{a_1}{a_2}\frac{\beta_1}{\beta_2}\frac{\alpha_2}{\alpha_1}\right)^{\gamma_1'} F^{\gamma_2'} \quad (18)$$

where:

$$\gamma_1' = \frac{1}{1 + \beta_1 - 1/\tau}\ ;\ \gamma_1' > 0$$

$$\gamma_2' = \frac{1 + \beta_2 - 1/\tau}{1 + \beta_1 - 1/\tau}\ ;\ \gamma_2' > 0$$

$$X = \left(\frac{a_3}{a_2}\frac{\beta_3}{\beta_2}\frac{\alpha_2}{\alpha_3}\right)^{\gamma_3'} F^{\gamma_4'} \tag{19}$$

where:

$$\gamma_3' = \frac{1}{1 + \beta_3 - 1/\tau}\ ;\ \gamma_3' > 0$$

$$\gamma_4' = \frac{1 + \beta_2 - 1/\tau}{1 + \beta_3 - 1/\tau}\ ;\ \gamma_4' > 0$$

$$\frac{P_M}{P_F} = \left(\frac{\alpha_1}{\alpha_2}\right)^{(1+\beta_1)\gamma_1'} \left(\frac{a_1}{a_2}\frac{\beta_1}{\beta_2}\right)^{-\gamma_1'/\tau} F^{(1-\gamma_2')/\tau} \tag{20}$$

$$\frac{P_X}{P_F} = \left(\frac{\alpha_3}{\alpha_2}\right)^{(1+\beta_3)\gamma_3'} \left(\frac{a_3}{a_2}\frac{\beta_3}{\beta_2}\right)^{-\gamma_3'/\tau} F^{(1-\gamma_4')/\tau} \tag{21}$$

Equations (18) and (13) are identical, as are (20) and (14). However, the implications of these equations are quite different in a three-commodity model from those of a two. This difference is best seen through the demand relationship underlying the three-good utility function. As chapter 4 notes, we distinguish partial elasticities of substitution between pairs of goods and a general elasticity of substitution for three. One measures the curvature of a point on a cross-section of the utility surface; the other measures curvature at a point on the surface in three dimensions.

Since the transformation contour exists in three-dimensional space, non-neutral changes require a specification of the relative variation in α_1, α_2, and α_3. If the α_1/α_2 ratio is increasing faster than α_3/α_2, then technical change favors F the most, then X, and then M. Clearly this example is an unrealistic case. Our attention will be directed to those cases in which the effects of technical change upon the amenity services of natural environments are negligible relative to those upon fabricated goods. The role we assign to X will determine the assumptions made regarding technical changes. Figure 15 illustrates a production possibility curve in three-dimensional space. If we assume ADB is the frontier in period t, then technical change causing a non-neutral pivoting about A requires specification of the relative movement of the M and X coordinates. AEC is one possibility. In this case, movement along the M axis exceeds that of X.

One convenient means of specifying the alternative possibilities is to describe the time paths of α_1/α_2 and α_3/α_2 as in equations (22) and (23):

$$\left(\frac{\alpha_1}{\alpha_2}\right)_t = m_1 e^{g_1 t} \tag{22}$$

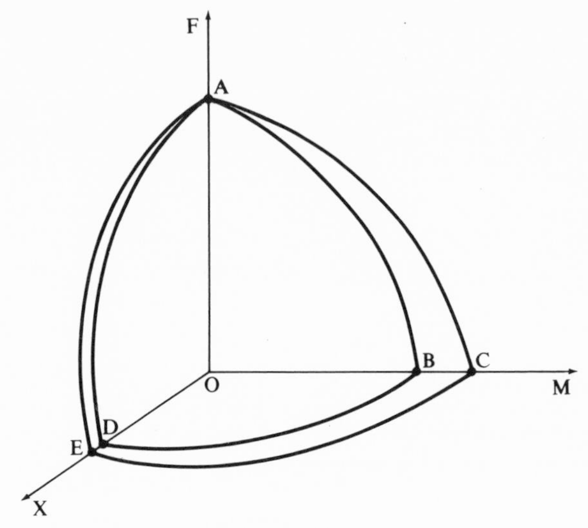

FIGURE 15.

$$\left(\frac{\alpha_3}{\alpha_2}\right)_t = m_2 e^{rt} \tag{23}$$

where: m_1, m_2 = constants

The conditions described by figure 15 require both g_1 and r to be less than zero and the absolute magnitude of g_1 to exceed that of r. Since we have assumed that technology will not favor the production of amenity services, the production frontier will be shifting outward in the XM plane. The absolute magnitude of r and g_1 determine whether it is parallel $(|r| = |g_1|)$, favors $M(|r| < |g_1|)$, or favors $X(|r| > |g_1|)$.

The distinction between the three-good and two-good models is clearest when reference is made to the determinants of relative price behavior. Accordingly, with our specified time functions for the effect of technical change upon society's production possibility locus, we can derive the three-good analog to equation (15). Differentiating (20), expressing it in terms of a rate of change, and then translating the parameters into terms of the demand elasticities, we have equation (24):

$$\frac{\dfrac{d\left(\dfrac{P_M}{P_F}\right)}{dt}}{\dfrac{P_M}{P_F}} = \frac{E_y(F)g_1}{E_y(F) - \dfrac{E_y(M) \cdot \sigma_{FM}}{\tau\sigma_{MX}(1+\beta_3)}} + \frac{E_y(F) - E_y(M)}{\tau E_y(F) - \dfrac{E_y(M)\sigma_{FM}}{\sigma_{MX}(1+\beta_3)}} \cdot \frac{\dot{F}}{F} \tag{24}$$

The important distinguishing feature between this expression and that of equation (15) is the denominator of each of the two terms on the right. The ratio of the partial elasticities of substitution in these two terms accounts for the extent to which the third good, X, functions as a substitute for either F or M. The magnitude of the term $[\sigma_{FM}/\sigma_{MX}(1 + \beta_3)]$ depends upon the degree to which F and M are substitutes relative to M and X. For example, if F and M are closer substitutes than M and X (all else equal), this relationship will tend to offset technical change, which is biased in favor of M. If the reverse is true, for a given τ, the first term in (24) will be closer in size to g_1. In economic terms, the rate of relative price change will approach the rate of technical advance.

The second term of (24) may reinforce or offset the effects of technical change. That is, the magnitude of the income elasticities of demand for amenity services and fabricated goods and the direction of change in the equilibrium consumption of amenity services will affect the net contribution of this term. If, for example, the income elasticity of demand for amenity services is assumed to exceed that of fabricated goods, and the equilibrium consumption of amenity services to be increasing over time, then this term will reinforce the effects to the first term to increase the price of F relative to M. However, these conclusions can be altered if the income elasticity relationship is not upheld with the equilibrium consumption of amenity services increasing, or if the consumption of amenity services is declining and $E_y(F) > E_y(M)$.

A number of factors must be considered before we define the direction of relative price movement. Table 4 summarizes some possible combinations of supply and demand conditions. For each case, we might also postulate alternative roles for X and develop further scenarios. However, these examples serve to illustrate the possibilities.

It may be useful at this juncture to stop and question the implications of the model in its present form. Equation (24) provides a direct link between relative prices and technical change while accounting for the nature of community demand and supply. Thus, if we are faced with assessing the extent to which a good's or service's price will appreciate over time, we need to answer the following questions:

(1) What is the relative luxury status of the good in question—i.e., how does $E_y(F)$ compare to $E_y(M)$?
(2) Are there convenient substitutes available for the goods whose relative prices we are concerned with—i.e., what is the magnitude of $\sigma_{FM}/\sigma_{MX}(1 + \beta_3)$?
(3) What is the direction and rate of technical advance—i.e., what is the magnitude of g_1?

Table 4. Relative Price Behavior in a Three-Good World: $(\dot{P}_M/P_F)/(P_M/P_F)$

Case	$E_y(F) > E_y(M)$	$E_y(F) = E_y(M)$	$E_y(F) < E_y(M)$
(1) $\tau \to 0$, $\frac{\dot{F}}{F} > 0$	$-\frac{E_y(F) - E_y(M)}{\frac{E_y(M)\sigma_{FM}}{\sigma_{MX}(1+\beta_3)}} \cdot \frac{\dot{F}}{F}$ (declining)	0 (no change)	$-\frac{E_y(F) - E_y(M)}{\frac{E_y(M)\sigma_{FM}}{\sigma_{MX}(1+\beta_3)}} \cdot \frac{\dot{F}}{F}$ (increasing)
(2) $\tau \to 0$, $\frac{\dot{F}}{F} = 0$	0 (no change)	0 (no change)	0 (no change)
(3) $\tau \to 0$, $\frac{\dot{F}}{F} < 0$	$-\frac{E_y(F) - E_y(M)}{\frac{E_y(M)\sigma_{FM}}{\sigma_{MX}(1+\beta_3)}} \cdot \frac{\dot{F}}{F}$ (increasing)	0 (no change)	$-\frac{E_y(F) - E_y(M)}{\frac{E_y(M)\sigma_{FM}}{\sigma_{MX}(1+\beta_3)}} \cdot \frac{\dot{F}}{F}$ (declining)
(4) $\tau \to -\infty$, $\frac{\dot{F}}{F} > 0$	g_1 (declining)	g_1 (declining)	g_1 (declining)
(5) $\tau \to -\infty$, $\frac{\dot{F}}{F} = 0$	g_1 (declining)	g_1 (declining)	g_1 (declining)
(6) $\tau \to -\infty$, $\frac{\dot{F}}{F} < 0$	g_1 (declining)	g_1 (declining)	g_1 (declining)

(4) Is it possible to reproduce the good in question? If so, how easily may we convert resources devoted to the production of other goods and services to the production of the good of interest—i.e., what is the magnitude of τ?

(5) Over time, are population and other exogenous forces causing the equilibrium consumption of the good of interest to increase—i.e., what is the direction of change in $\dot{F}/F$?

With answers to these questions, we can describe explicitly the rate of change in relative prices.

The essential point of our model has been to link demand characteristics and supply properties in order to determine explicitly the primary influences upon relative price behavior. The specific problems we wish to apply the model to are the amenity services emanating from natural endowments. Characteristically, we cannot augment supply from a given area without altering the quality of each individual's experience. Such services are thus considered independent of the forces of technical change. This inability to increase their effective supply does not, however, make the prices independent of the consequences of technical change. And it is this conclusion that is of fundamental importance to the evaluation of the economic value of scarce natural environments.

SUMMARY

This chapter has described two- and three-good models in which disembodied technical change has been assumed to affect certain subsets of the goods differentially. Externalities have been assumed nonexistent for analytical convenience. Several overall findings may be summarized as follows:

(1) Homogeneous utility specifications for community preferences imply that substitution effects of the traditional character are the only demand determinant of relative price behavior. This specification assures the unitary income elasticity of demand for each good. The rate of relative price change for a given rate of technical change is thus a function of the community's ability to accept one of the two goods as a substitute for the other, relative to its ability to transform its allocation of resources so as to alter the produced outputs.

(2) Nonhomogeneous two-good utility functions allow for both income and substitution effects as determinants of relative price behavior, in addition to the supply effects. However, the substitution options are limited to the two goods under study.

(3) The introduction of a third good with a nonhomogeneous utility specification requires taking account of the extent to which it can function as a substitute for the other goods. Moreover, technical change must be specified in terms of its relative effect upon the third good as well.

REFERENCES

Arrow, K. J., et al. 1961. "Capital-Labor Substitution and Economic Efficiency." *Review of Economics and Statistics* 43: 225–250.

Barnett, H. J., and C. Morse. 1963. *Scarcity and Growth*. Baltimore: The Johns Hopkins Press.

Baumol, W. J. 1967. "Macroeconomics of Unbalanced Growth: The Anatomy of Urban Crisis." *American Economic Review* 57: 415–426.

Hicks, J. R., and R. G. D. Allen. 1934. "A Reconsideration of the Theory of Value." *Economica* 1: 52–77, 196–219.

Krutilla, J. V. 1967. "Conservation Reconsidered." *American Economic Review* 57: 777–786.

Mukerji, V. 1963. "A Generalized SMAC Function with Constant Ratios of Elasticity of Substitution." *Review of Economic Studies* 30: 233–236.

Powell, A. A., and F. H. Gruen. 1969. "The Constant Elasticity of Transformation Production Frontier and Linear Supply System." *International Economic Review* 9: 315–328.

Smith, V. K. 1972. "The Effects of Technological Change on Different Uses of Environmental Resources." In *Natural Environments: Studies in Theoretical and Applied Analysis*, ed. J. V. Krutilla. Baltimore: The Johns Hopkins University Press for Resources for the Future, Inc.

Worcester, D. A., Jr. 1968. "Macroeconomics of Unbalanced Growth: A Comment." *American Economic Review* 58: 886–893.

chapter six Intertemporal Externalities, Technical Change, and the Price of Amenity Services

Externalities are interactions that bring costs or benefits to individuals without their consent and without market-determined compensations. Considered in this light, irreversibilities are an important type of intertemporal externality. Irreversibilities occur when allocations of natural environments necessarily preclude the provision of alternative service flows in the future. Consider the case of open-pit mining. Once a natural area has been designated for this kind of extractive activity, it cannot easily be returned to its original state. Although the mines may be refilled and the surrounding area regenerated, it is unreasonable to suggest that unique geomorphological wonders can be exactly replicated in the process. The purpose of this chapter is to adapt the model presented in the preceding chapter so that irreversibilities can be accommodated in the analysis.

We first consider this kind of intertemporal externality and its modeling. The second section reviews the general equilibrium model developed in the preceding chapter and the effect of irreversibilities upon the related prices of amenity services. The last section summarizes the discussion.

STATIC VERSUS INTERTEMPORAL EXTERNALITIES

The analysis of externalities has traditionally been a static one, in which the effects of one individual's actions on others are viewed as contemporaneous phenomena. There is, however, nothing in conventional definitions to prevent their being broadened to include intertemporal effects as well. Thus, we might define externalities as taking place whenever one set of individuals' actions affect another set and formal mechanisms for accounting for these effects are absent. For example, the activities of a steel mill will have many effects upon the residents in some neighborhood of the plant. These operations provide a demand for labor services that can be exchanged in the market. They also result in air pollution. Since the air mantle is collectively owned, the disposal of the residues by one economic agent cannot be controlled through the market. Hence those individuals in the same locality (the specific definition of which will depend on meteorological conditions) as the mill cannot rely on the market to diminish air pollution caused by the steel mill's activities. Similarly, the actions of one consumer may impose externalities on other consumers or producers. If I wish to swim in a river that meets minimum health standards and a firm wishes to expel its industrial residues into the same river, my actions (if my wishes can be enforced) will have implications for the firm. Where property rights are not specifically assigned, whoever can put the services of the common property resource to his own uses may generate externalities.

The same type of interaction can be envisioned with economic agents over time. Recent work by Krutilla (1967), Fisher, Krutilla, and Cicchetti (1972), and Arrow and Fisher (1974) suggests that these intertemporal externalities constitute an important but often excluded component of the set of external effects. An individual's or, more generally, society's decisions at any time impose constraints not only upon others at that time, but also upon society in the future. Since past decisions are beyond the control of society in the present, every allocative decision must be scrutinized to assess its intertemporal effects. Such careful review is particularly important when the allocation will be irreversible. For example, transforming an environmental resource from a pristine wilderness to a mineral-producing facility is, by the technical nature of the mineral production, irreversible. If, in addition to the irreversibility of such actions, these environmental resources cannot be produced by man, then their present and future values in alternative uses must be carefully analyzed before efficient decisions can be made.

The basic premise of the model developed in chapter 5 is that economic activities may be classified according to the effect technology has upon them. As we noted, amenity services are derived from environmental resources and, by their very nature, are *not* capable of technical augmentation. The production processes of fabricated or manufactured goods, on the other hand, are subject to technical improvements, and therefore larger supplies may be available over time. The model in chapter 5 measured the effects of such asymmetrical technical change upon the relative prices of amenity services and the manufactured goods. It did *not*, however, account for the effects of irreversibilities. That is, while we have indicated that certain uses of the direct services of an environmental resource do not affect the ability of that resource to provide service flows in the future, other uses do. Thus, if the direct services of environmental resources are used to produce fabricated goods in the present, there will be fewer of these services available in the future for either amenity services or fabricated goods. The allocation is then an irreversible one and herein lies the source of the intertemporal externalities we shall consider.

Once we allocate our resources to the production of any good, we cannot (in general) reverse the action with those same resources. Once I have decided to plant corn in a field for the season in question, and not clover, it is not possible to reverse the action. In the next season, during the appropriate planting period, I can have my clover. But consider another case: I have the choice of planting corn or selling the land to a developer to construct homes. If I do the latter, the extent of my commitment of the resource is more extreme. In the next period, to change my decision would be extremely costly. I would need to clear the land of all homes and to seed. It is possible to think of progressively more costly actions, where the extent to which the commitment is reversible depends on the costs of undoing what has been done. Irreversibility is the extreme case, in which we cannot reproduce the resource in its original state. We may be able to produce a close replica, but since authenticity is one of the attributes providing the value, the cost of reversing the action in this case may as well be considered infinite. It is in this sense that our models and the intertemporal externalities considered will differ from conventional treatments of the consequences of alternative resource allocations.

In order to integrate irreversibilities into our model, we must look at their effect upon the locus of production possibilities. Society's consumption choices in t affect the dimension of the Edgeworth-Bowley box in $t + 1$. Thus we should be able to trace the effects upon the commodities or services available for consumption in $t + 1$. Consider the model developed in chapter 3 for deriving the production possibility frontier, using Cobb-Douglas production functions. If we designate the hypothetical factor inputs (i.e., X_1 and X_2) the services of labor and the direct services of an environmental resource, respectively, and specify the outputs as fabricated goods and amenity services, then we can illustrate the effects of both asymmetrical technical change and irreversibilities.

Recall that perfect competitition in the factor markets was assumed so that the equilibrium hiring rule was used to derive equations for the coordinates of the transformation locus (see chapter 3, equations [16] and [17]). These are rewritten in (1) and (2) below, where M and F replace A and B, respectively:

$$\frac{F}{\bar{X}_1} = as\left(\frac{\beta}{1-\beta}\frac{\bar{Y}}{r}\right)^{1-\alpha} \tag{1}$$

$$\frac{M}{\bar{X}_1} = b(1-s)\left(\frac{\alpha}{1-\alpha}\frac{\bar{Y}}{r}\right)^{1-\beta} \tag{2}$$

where: $r = \dfrac{\alpha}{1-\alpha} + s\dfrac{\beta-\alpha}{1-\beta-\alpha+\alpha\beta}$

$$s = \frac{X_{1F}}{\bar{X}_1}$$

$$\bar{Y} = \frac{\bar{X}_2}{\bar{X}_1}$$

One way of modeling the asymmetric technical change described earlier is to assume that over time b in (2) is increasing while a is not. The production possibility locus will thus be shifting out along the M axis, but not along F. For simplicity, assume that it may be described as in equation (3):

$$b_t = b_0 e^{\rho t} \; ; \rho > 0 \tag{3}$$

Technical change is assumed to be progressing at a rate, ρ, independent of the resources used in the sector. Moreover, it affects both factor inputs equally. One simple means of modeling the effect of the irreversibilities previously described is to assume that the depletion of the current service flow from the environmental resource (i.e., X_{2_t}) is related to the previous period's output of the manufactured good, M.

$$\bar{X}_{2_t} = X_{20} e^{-\theta_t t} \tag{4}$$

where: $\theta_t = f(M_{t-1})$; $\dfrac{\partial \theta_t}{\partial M_{t-1}} > 0$

Substituting from (3) and (4) into (1), we have

$$\frac{M_t}{\bar{X}_1} = (b_0 e^{\rho t})(1 - s)(\varphi X_{20} e^{-\theta_t t})^{1-\beta} \tag{5}$$

where: $\varphi = \dfrac{\alpha}{1-\alpha} \cdot \dfrac{1}{\bar{X}_1 r}$

Our previous analysis has indicated that technical change affecting the production of M results in a shift in the transformation curve along the M axis. In figure 16 the shift from AC to AB illustrates the effect. Irreversibilities as modeled above reduce the locus to DG, so that for some points (i.e., RG) the frontier is outside the old locus, whereas for others (i.e., DR), it falls inside the previous period's production set. In order to assess whether or not society is better off, one must impose a utility function upon the diagram and thereby select a point on DG.

If we assume θ_t to be invariant at any specific time, then equation (6) indicates the condition when the production possibility frontier of period $t + 1$ will *not* lie outside that of t. Thus irreversibilities will offset the effects of technical advance.

$$\rho = (1 - \beta)\theta_t \tag{6}$$

If $g = (1 - \beta)\theta_t$, then the rate at which the contribution to the output of M lost as a result of the reduction in the stock of environmental resources just balances the rate at which innovation is improving pro-

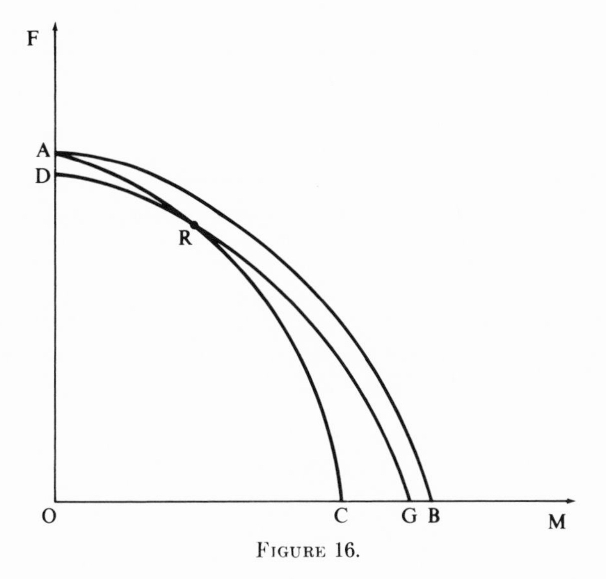

FIGURE 16.

ductivity, and the transformation locus remains stationary on the M coordinate.

The preceding model incorporates our basic idea of the effect of irreversibilities. It is, however, constructed in terms of production functions, and the model developed in the preceding chapter specifies a functional form for the production possibility curve, thereby subsuming a class of production relationships. It is relatively easy to allow for technical change within the *CET* framework developed in chapter 5. Moreover, our approach is completely consistent with that which a production function analysis would dictate.

In order to maintain the same theoretical constructions, we shall specify a simple shifting in the transformation curve along the vertical axis to account for the effect of irreversibilities. In figure 17, AA' is the curve prior to the decision; the origin for AA' shifts from O to O', so that the new transformation curve, after irreversible allocation decisions, is TT'. The amount of the vertical shift is a function of the previous period's choice of goods and services. This format implicitly assumes divisibility. That is, our model assumes that the broad choices of preservation versus development may be partitioned into smaller segments. For example, part of a national forest may be clear-cut, thereby precluding some fraction of the maximum quantity of amenity services that might have been enjoyed. Although the nature of the development in

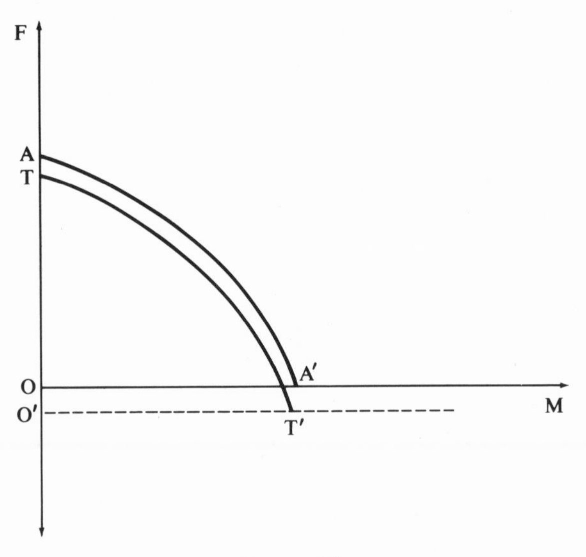

FIGURE 17.

some specific cases does not allow for such divisibility, within the economy as a whole there may be scope for this assumption.

Our formulation of the impact of developmental decisions is such that the quantity of amenity services foregone in period t, F_t^*, is the direct result of past decisions for the consumption of fabricated goods. Thus current trade-offs with fabricated commodities are not reflected in the irreversibility effect. Equation (7) specifies the cumulative function:

$$F_t^* = g(M_{t-1}, M_{t-2}, M_{t-3}, \cdots, M_{t-n}) \tag{7}$$

where: $\dfrac{\partial g}{\partial M_{t-i}} > 0$ for all i

Figure 18 demonstrates the joint effects of technical change favoring M and irreversibilities on the equilibrium output mix and relative prices with a specified community utility function. The curves labeled CIC_1 and CIC_2 represent two levels of community welfare and the corresponding sets of M (manufactured goods) and F (amenity services) to which society is indifferent at each level of satisfaction.

Assume that in period t the commodity mix chosen is OS of F and OR of M. Given the autonomous nature of technical change in our model

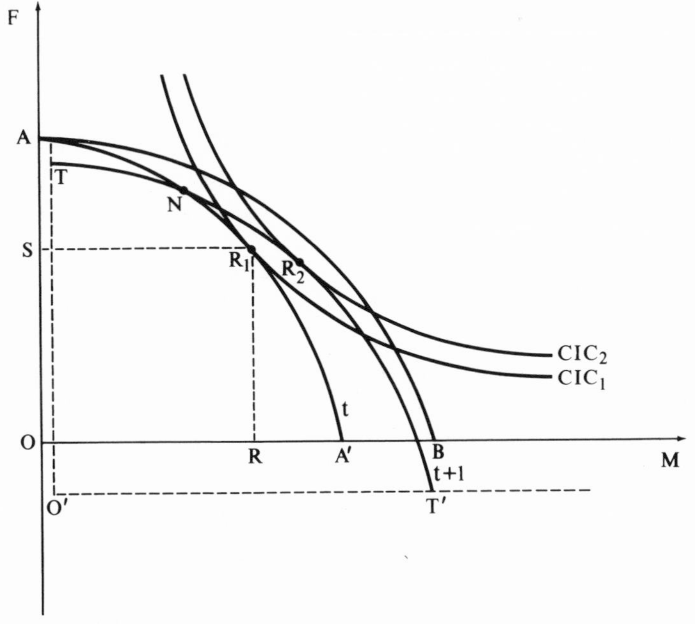

FIGURE 18.

and the relationship defined in (7), we can specify the equilibrium commodity mix and price ratio in period $t + 1$. Suppose that the production and consumption of OR of the fabricated good in period t necessarily reduces the quantity of amenity service by AT for the next period; then, in the absence of technical change, the frontier, AA', would shift down along the vertical axis. However, technology has been postulated to favor M. Consequently the production possibility frontier in $t + 1$ is TT', where the movement from AA' to AB is the result of technology, and the shift of AB down AT is the result of the previous periods' choice of M.

Clearly this construction is solely for diagrammatic simplicity. Analysis in terms of production functions is more thorough. Since our primary concern is with the outcome of these changes, it is possible to be satisfied with a consistent method for accounting for them. Thus our model will depict technological change as "one kind of shift" in the transformation curve and permanent destruction of X_2 as another.

The new equilibrium point in figure 18 is given by R_2. The relative price of M to F has declined as a result of the movement from the old equilibrium at R_1 to the new one. The objective of the model that follows will be to take explicit account of these causal determinants of relative prices.

Before we proceed to the model, however, the issue of intertemporal externalities and nonconvexities should be considered. Baumol and Bradford (1972) have suggested that sufficiently serious production externalities are likely to cause nonconvexities in the social production set. The reasoning underlying their statement is straightforward. Suppose we have a society in which two firms each produce one good. If the production of one of these goods generates industrial residues or wastes, and these are dispersed in such a way that they affect the other firm, then production externalities are present. Assume further, for simplicity, that the receiving firm does not impose reciprocal effects upon the polluter.

If society chooses to produce all of the good supplied by the polluting firm, then the equilibrium position will be on the same terminus as it would have been in the absence of the externalities. Alternatively, if the commodity mix included only the output of the nonpolluting firm, the endpoint would also be unchanged. However, at every point between the endpoints of the transformation curve, producing more output by the polluter not only requires a larger share of the presumed fixed factor supplies, but also, and more importantly, inhibits the ability of the other firm to produce its output at any level of resource usage. There are thus more than the usual increasing marginal costs at work when externalities are present. Although the extent of nonconvexity will be related to a number of factors (e.g., severity of the externality, the nature of the elasticities of substitution between the factors in production, etc.), it is safe to say that the traditional concave shape of the curve will not be maintained.

Baumol and Bradford (1972) note further that:

> There is only limited comfort to be derived from the knowledge that a sufficiently ingenious use of Pigouvian taxes can keep a competitive economy at any desired point that is technologically efficient so long as detrimental externalities are the only source of non-convexity. Pigouvian taxes cannot change the shapes of the technological relationships in the economy, and hence cannot remove the problems of evaluation of efficiency which non-convexity introduces. [p. 172]

Their view can hold only in a static framework, however. Once technical change and the innovation underlying it are seen to be processes endogenous to an economic system, our results do not look as bleak. That is, innovation responds to perceived scarcity, and thus Pigouvian taxes provide the signals that will direct technological innovation to ameliorate the nonconvexities (see Smith 1974).

This discussion relates only marginally to the issue at hand: whether intertemporal externalities, like static production externalities, introduce nonconvexities. The answer lies in the definition of the externalities in question. Intertemporal externalities affect the production possibility frontier through their influence on the resource base available for production in the future. The kind of externality we are dealing with thus differs from the kind Baumol and Bradford consider. In our case, a movement along either axis (which changes the selected commodity mix) does not affect the production of the other commodity aside from altering the distribution of the fixed stocks of factor inputs. Once a choice from the locus of production possibilities is made, we preempt certain future selections by using a subset of the resources available to the present period. Thus certain consumption expansion paths are ruled out because it is technically impossible to achieve them. It can, however, be argued that intertemporal externalities do *not* introduce nonconvexities in any period's social production set.

REVISED THREE-GOOD MODEL WITH NONHOMOGENEOUS SPECIFICATION

The preceding discussion was necessary so that irreversibilities might be introduced into the framework developed in earlier chapters. We have suggested that it does not seriously alter the nature of irreversibilities to conceive of them as shifting the production possibility frontier along the vertical axis (when F is measured along it). Thus, in figure 19, a change from transformation curve $TT'T''$ to $RR'R''$ is the result of irreversibilities and technological change favoring M and X. Our objective is to alter the model developed in chapter 5 so as to allow these kinds of movements. The utility function stated by equation (16) in chapter 5 will remain the index of community preferences for our model. However, the transformation curve described by (17) will be replaced by:

$$\alpha_1 M^{1-1/\tau} + \alpha_2(F - F^*)^{1-1/\tau} + \alpha_3 X^{1-1/\tau} = k \tag{8}$$

This function will change through the relative values of α_1, α_2, and α_3 over time as a result of technical change (see equations [22] and [23] in chapter 5), and F^* reflects the effects of irreversible resource allocations, as in equation (7). Recall that, in this model, comparative static equilibrium is achieved when one of the CIC surfaces is tangent to the production possibility locus, so that the marginal rates of substitution for all pairs of goods are equal to the corresponding marginal rates of transformation. Substituting the expressions for the rate of technical change given in chapter 5 into the equations derived for equilibrium relative prices with the new transformation curve, and differentiating logarith-

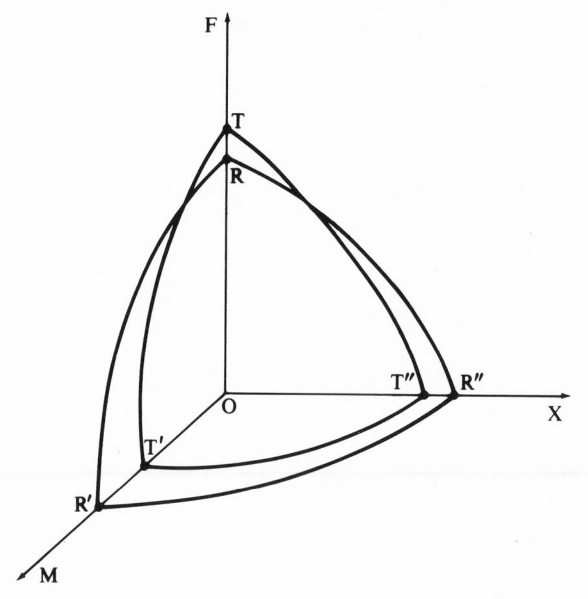

FIGURE 19.

mically with respect to time, we can derive an expression for the rate of relative price change, as we saw in chapter 5. Moreover, making use of the Hicks-Allen equations, we can replace the parameters of the utility function with demand elasticities, providing an expression comparable to (24) in the preceding chapter:

$$\frac{\frac{d(P_M/P_F)}{dt}}{P_M/P_F} = \frac{E_y(F)g_1}{E_y(F) - \frac{E_y(M)\cdot\sigma_{FM}}{\tau\sigma_{MX}(1+\beta_3)}} + \frac{E_y(F) - E_y(M)}{\tau E_y(F) - \frac{E_y(M)\sigma_{FM}}{\sigma_{MX}(1+\beta_3)}} \cdot \frac{\dot{F}}{F - F^*} \tag{9}$$

This expression looks very much like its predecessor in chapter 5. The only difference is in the second term. Here the growth in equilibrium consumption of amenity services is considered relative to a new base, which allows for the increase in the *absolute* scarcity that irreversibilities cause for amenity services. The postulated path of technical change means that amenity services feel the full adverse effects of the irreversibilities.

We would in fact want to adjust the rates of technical advance to reflect the effect of reduction in the supply of the direct services of natural environments as a result of the intertemporal effects. Clearly the extent of such an adjustment will depend upon both the nature of the specific production processes involved and the availability of substitutes in production for the services of natural environments. By maintaining the rate of advance at g_1, our analysis may be somewhat misleading. Consequently, g_1 in (9) ought to be replaced by a g_1^*, which is less than or equal to g_1. The effect of intertemporal externalities upon the output of fabricated goods may not be severe. For example, in our earlier discussion using Cobb-Douglas production function, the net movement (assuming the F terminus in a two-good world is stationary) depends not only upon the rate of depletion in the direct services of the environmental resource (i.e., θ), but also upon the elasticity of productivity of these resources in the production of fabricated goods. Accordingly, for the two-good case, g_1^* might be expressed as:

$$g_1^* = g_1 - (1 - \beta)\theta \tag{10}$$

In order to maintain the g_1 in (9), we are assuming $(1 - \beta)$ is very small. That is, the elasticity of productivity of the direct services of an environmental resource in the production of fabricated goods is small.

The validity of this assumption is clearly an empirical question. Moreover, the assumption of unitary elasticity of substitution, which underlies the Cobb-Douglas production function specification, must also be empirically tested before a judgment can be made.

Since the model we have developed originates with a production transformation function rather than with individual production functions, the effects of irreversibilities on g_1 cannot be well specified. The preceding discussion is meant to indicate the kinds of considerations likely to be important in such an analysis.

Further assumptions should also be noted. The expression for the rate of relative price change in equation (9) assumes that the preferences of future generations will be the same as those of the present. If there is reason to assume that tastes will change over time, through a learning-by-doing process or in response to institutional factors, then such changes will not be reflected in the relationship in equation (9). It is, however, possible to integrate these changes into the analysis by assuming that the parameters of the community utility function will move in one path or another over time. Such change is *not* essential to the measurement of the effects of the intertemporal externality we have chosen to study. Irreversibilities are the result of the characteristics of certain resources and our ability to reproduce them. These externalities are important because the services provided by the resources are valuable to a reasonably large subset of the community. In the absence of changing preferences, it is possible to show that decisions efficient from a static viewpoint may result in long-run welfare losses. However, my concern here is not with establishing criteria to prevent these decisions. The models have been directed instead toward explaining the implications of decisions, made on the basis of static efficiency, for relative prices.

SUMMARY

When the model developed in chapter 5 is generalized to allow for intertemporal externalities, the overall conclusions of that chapter are not affected. However, the severity of the effects, in terms of the rate of appreciation of the relative price of amenity services, is more pronounced. Several additional observations should be noted:

(1) The conventional definition of externalities can be broadened to include the effect of an individual's activities in the present upon the value of a production or a consumption function of his contemporaries, as well as those individuals in the future, who might be affected.

(2) After examining the implications of irreversibilities for the social production set using the two-product, Cobb-Douglas production function model developed in chapter 3, we described a procedure for integrating them into the framework of the model developed from a transformation curve. The transformation curve is translated along the vertical axis (i.e., that axis where the quantity of amenity services is measured) according to past decisions for fabricated goods.

(3) The revised model, incorporating both technical change and irreversibilities, indicates that the rate of relative price appreciation will be accelerated by the increase in the absolute scarcity that irreversibilities necessarily imply.

REFERENCES

Arrow, K. J., and A. C. Fisher. 1974. "Environmental Preservation, Uncertainty, and Irreversibility." *Quarterly Journal of Economics* (forthcoming).

Baumol, W. J., and D. F. Bradford. 1972. "Detrimental Externalities and Nonconvexity of the Production Set." *Economica* 39: 160–176.

Fisher, A. C., J. V. Krutilla, and C. J. Cicchetti. 1972. "The Economics of Environmental Preservation: A Theoretical and Empirical Analysis." *American Economic Review* 62: 605–619.

Krutilla, J. V. 1967. "Conservation Reconsidered." *American Economic Review* 57: 777–786.

Krutilla, J. V., and C. J. Cicchetti. 1973. "Evaluating Benefits of Environmental Resources with Special Application to the Hells Canyon." *Natural Resources Journal* 12: 1–29.

Mishan, E. J. 1971. "The Postwar Literature on Externalities: An Interpretative Essay." *Journal of Economic Literature* 9: 1–28.

Smith, V. K. 1974. "Detrimental Externalities, Nonconvexities, and Technical Change." Mimeo.

chapter seven Conclusions

The models developed in the preceding chapters have sought to develop a systematic framework within which both demand and supply influences upon the pattern of relative price changes might be analyzed. The purposes of this modeling effort were twofold: (1) to develop comparative static analysis so as to isolate the demand and supply forces important to the market determination of the opportunity costs of resource allocations, and (2) to formulate a structure that allows one to bound the parameters of Fisher, Krutilla, and Cicchetti's (1972) model for resource allocation decisions through benefit-cost analysis, given the presence of technologically induced changes in relative prices.

Traditionally, benefit-cost analysis has attempted to "adjust" its estimates for potential biases due to inflationary or deflationary movements in the general price level. However, changes in the relative prices of goods and services due to such non-project-related forces as asymmetric technological change have not been taken into account. The models constructed here clearly indicate that the importance of such factors as asymmetric technical change and irreversibilities depends on the structure of community demand, as well as on supply considerations.

The framework for our analysis has been a simple general equilibrium model in which each side of the market (i.e., demand and supply) is represented with summary functions. Such specifications imply some behavioral assumptions, which we have discussed in detail. On the supply side of the market, we need to assume that when individuals produce their amenity services there are few substitutes for the services provided by the natural environment. This statement is a formal means of discussing one's ability to enjoy a wilderness experience without the natural surroundings that are generally part and parcel of the wilderness scene. Although one can ride in a federal paddock and/or walk on a city street, even the nonwilderness enthusiast will grant that such experiences are wholly different in character from backpacking or riding in a remote, undisturbed wilderness setting. These are the service flows we are referring to in our production functions.

If this limited substitutability between the services of an environmental resource, such as a wilderness, and other factors is recognized, then the character of the environmental services will not appreciably affect our summary function. Why is this important? Our exercise has been deliberately simplified. We have specified the forms of our summary functions, so that an explicit link can be made between the rate at which technology is assumed to be progressing and the behavior of relative prices. Although this exercise does help to conceptualize the problem, it has no great practical significance unless the models approximate real-world behavior. By inquiring into the assumptions required to make the model a reasonable representation of behavior, we are attempting to specify the degree of approximation involved.

The services of a natural environment have many of the characteristics of "publicness" in supply in that over certain levels of use, providing such services to an additional person for the purpose of recreation or other amenity services, in the absence of congestion, has negligible marginal cost. If, however, these services are used for extractive purposes leading to the production of manufactured goods, and we are willing to ignore future generations, then ownership can be readily assigned, and the marginal cost function has a more traditional shape. Hence, depending upon which sector or use is required of the resource, there may be a discrepancy in its "price." Such discrepancies in factor markets may lead to distortions in the transformation curve describing a society's options. However, the extent of the distortion is a function of the ability to substitute for such factors in the respective production processes. Hence it would seem that for our case, in spite of factor market imperfections, we can approximate the traditional neoclassical transformation contour.

On the demand side of the market, general equilibrium analysis, in the past, has proceeded rather easily by specifying a community indifference mapping to describe the preferences of all members of the society. Although this tool remains in use in current models of behavior in international trade theory, it has been otherwise ignored in more sophisticated analysis of general equilibrium behavior, primarily because it cannot be derived from individual preference functions. Our purposes do not require placing the individual in perspective with the whole. A general description of the whole community preference structure will suffice, so long as the demand parameters (i.e., price, cross, and income elasticities) it implies conform with what we know from empirical studies of behavior.

In the absence of intertemporal externalities, or factors that link resource allocations at one time with what is feasible at another, we can show that the considerations in determining the magnitude of the effect of asymmetric technical change upon relative price behavior are: (1) the income elasticity of demand for amenity services relative to that of fabricated or manufactured goods, (2) the availability of good substitutes for these amenity services, and (3) the ability of the economy to transform its resources so as to produce alternative commodity mixes.

The first of these considerations is straightforward. Our analysis suggests that a relatively larger percent increase in demand for amenity services than that for manufactured goods will result from a given increase in income. Hence these amenity services are termed income-elastic. The second consideration also refers to the character of demand; the community feels there are few, if any, good substitutes for these amenity services. An important distinction has been made between our ability to substitute for the services of an environmental resource in the production of these amenity services and the degree to which the community's taste patterns allow for substitution for amenity services in the process of consumption decision making. The second condition refers to substitution in terms of demand rather than production.

The last factor refers to the character of the process by which amenity services and manufactured goods are supplied. Movement along a production possibility frontier evaluates the alternative, efficient output combinations available to society. Such movement necessarily defines how readily our resource base may be transformed into one or the other of the goods under consideration. A constant rate of transformation implies that at any level of output of either commodity, we can get some constant fraction of the amount of the other good given up. Generally, we suggest that the law of diminishing returns is operative, so that as we move toward additional amounts of fabricated goods, we must forego ever-increasing quantities of amenity services. More specifically, we refer to the extent to which diminishing returns are operative in each of the sectors of our model.

The intertemporal externalities we have considered result from certain allocations of our environmental resources and contribute to the relative scarcity of the amenity services precipitated by asymmetric technical change and the character of the demand structure. Consequently, when these externalities are entered into the models, the increasing relative value of the amenity services is further accentuated. Thus, the value of preserved natural environments relative to the same resources in a developed state will correspondingly increase.

Throughout, we have postulated that technical change is autonomously given and favors the production of fabricated goods. This assumption generally follows established empirical results, particularly in the extractive industries. Barnett and Morse (1963), for example, have suggested that technological change has compensated for the depletion of higher-quality natural resource stocks by permitting the use of lower-quality resources without real cost increases. Other empirical studies, such as those reported in Brown (1967), find that a fairly substantial fraction of the measured growth in productivity can be attributed to technical change alone. Thus, with the exception of "cranks,"[1] economists have generally been most optimistic in their anticipations for the future.

There is little unambiguous evidence to support either this view or a more pessimistic one. In fact, there are some limited indications of uneasiness with our measurements of the unexplained portion of productivity growth traditionally associated with technical change.[2] Our national and sectoral accounts do not record the consumption of nonpriced common property resources in the production process. These resources

[1] Robinson (1972) noted in her Richard T. Ely lecture that "a sure sign of crisis is the prevalence of cranks. It is characteristic of a crisis in theory that cranks get a hearing from the public which orthodoxy is failing to satisfy. . . . The cranks are to be preferred to the orthodox because they see that there is a problem" (pp. 8–9).

[2] Jorgenson and Griliches (1967) contend that "all" of technical change's effect upon productivity growth can be explained by properly taking into account the aggregation and measurement errors in prices and quantities of the inputs and outputs.

serve not only as receptacles for industrial residuals, but also as basic inputs to the production process (see Russell 1973, for example). The market's failure to account for these costs prevents our netting out their contribution to the value of the produced output. Consequently, changes in their use over time and the associated deterioration in the quality of our environmental resources, such as air and water, are likely to offset some fraction of the productivity growth previously attributed to technical advance.

Recent analytical modeling and corresponding empirical results have supported the view that technical change is endogenously determined. That is, inventions are developed in response to perceived needs. This does not, of course, deny that there may be substantial random influences at work in the process of such invention. However, firms recognize that research and development activities generate both process and product changes. Once inventions are generated, the process by which they are implemented is designated an innovation in the productive technology. The set of the inventions has been screened with consideration given to the feasibility of implementation of these changes. The perspective for such an evaluation with process innovation will generally be cost saving, or potentially cost saving, in response to perceived factor price increases.

Once the innovation has had time to affect the production of output, it is termed a technological change. This genealogy is a rough-and-ready sketch of the modeling that generally underlies the response of society's technology set to incentives in our economic system. Where organized markets function, resource scarcities will be perceived through the prices established in the marketplace, and the invention-innovation process, over time, should respond. Common property resources, on the other hand, do not exchange in functioning markets; as a result, our private exchange process has ignored the increase in their costs due to scarcities. Moreover, technical change has developed production processes that use these resources extensively, both as factor inputs and as receptacles for industrial wastes. Once the real costs can be imposed upon those using such resources, past experience suggests that the innovation process will respond, reducing the deleterious effects of existing technology.

Taking account of the use of common property resources and their social costs is likely to alter our perceptions of the impact of technological change. Given the uncertain nature of the invention-innovation process, it is not unreasonable to suggest that, for some industries, pricing the services of common property resources may induce technologies that represent overall improvements (see Smith 1972). That is, one suspects that for some industries, production processes that provide less usable output with a given resource base but substantially fewer generated residuals may be selected as a profit-maximizing technology. Without a price on the disposal of these industrial residues, such a change would not be considered an improvement by the firm.

General conclusions to these questions will not be available on an aggregative level. The Measure of Economic Welfare measure devised by Nordhaus and Tobin (1972), though instructive, is unlikely to provide much insight into the real impact of technology on social productivity. Though they have attempted to account for the nonmaterial disamenities that have been accruing as costs to our economy, such adjustments at an aggregate level are necessarily *ad hoc.*

Nonetheless, research into the real effects of technology is needed. We have postulated that technical advance proceeds at a given rate, favoring the production of fabricated goods, but specific attention must be given to the magnitude of the rate in order to gauge the seriousness of the relative price movements predicted by our models. That is, once the social costs of productivity increases are accounted for, a smaller "gain" in relative productivity in the manufactured goods sector is likely. *A priori* analysis suggests that appropriate pricing of the services of common property resources will reduce the generation of industrial residuals so that pollution will be diminished to efficient levels. We do not, however, have any idea of the long-term effect of such pricing policy. The optimistic view holds that technology will be induced to solve all our problems.

At the other extreme, however, the changes in production possibilities we have described as due to technology may be simply the substitution of the services of common property resources for priced factors. Thus, instead of increasing our productive ability at a given level of resource usage, we are using more of one resource, previously nonpriced. Under this view of the world, we are trading off one resource use for another by technical change with little or no gain.

I am not suggesting that either of these two views is correct. At present, we have little or no basis on which to discriminate between them. Earlier research directed toward measuring technical change does not help us, since only the priced factors were considered. Once these analyses are reconstructed to include all factors, both priced and nonpriced, then there will be some information available for a judgment.

In the absence of this information, a reasoned guess would hold that our technical gains have probably been more limited than conventionally believed, when all their implications are accounted for. Hence, in terms of the models developed in this monograph, the effects of asymmetric technical change may be less important than the effects of irreversibilities. While the effects are intertwined in observed behavior, the models we have presented analytically partition these two factors, so that irreversibilities are those intertemporal externalities resulting from past consumption decisions which serve to reduce the stock of available services from environmental resources for all future generations. Thus, in an absolute sense, the level of amenity services available is reduced over time. Asymmetric technical change increases the community's ability to provide manufactured goods, leaving its ability to supply amenities unaffected. Thus, in relative terms, the amenity services become more scarce.

Public expenditure decisions must be made in the presence of both forces; the models developed in this monograph systematically document the demand and supply conditions important to this decision framework. What yet remains to be done is more careful evaluation of technical change, taking account of the unpriced consumption of common property resources.

REFERENCES

Barnett, H. J., and C. Morse. 1963. *Scarcity and Growth*. Baltimore: The Johns Hopkins Press.

Brown, M., ed. 1967. *The Theory and Empirical Analysis of Production*. New York: National Bureau of Economic Research.

Cicchetti, C. J., and V. K. Smith. 1973a. "Congestion, Quality Deterioration, and Optimal Use: Wilderness Recreation in the Spanish Peaks Primitive Area." *Social Science Research* 2: 15–30.

———. 1973b. "The Measurement of Individual Congestion Costs: An Application to Wilderness Recreation." Mimeo.

Fisher, A. C., and J. V. Krutilla. 1972. "Determination of Optimal Capacity of Resource Based Recreation Facilities." In *Natural Environments: Studies in Theoretical and Applied Analysis*, ed. J. V. Krutilla. Baltimore: The Johns Hopkins University Press for Resources for the Future, Inc.

Fisher, A. C., J. V. Krutilla, and C. J. Cicchetti. 1972. "The Economics of Environmental Preservation: A Theoretical and Empirical Analysis." *American Economic Review* 62: 605–619.

Jorgenson, D., and Z. Griliches. 1967. "The Explanation of Productivity Change." *Review of Economic Studies* 34: 249–283.

Kamien, M. I., and N. L. Schwartz. 1968. "Optimal 'Induced' Technical Change." *Econometrica* 36: 1–17.

Nordhaus, W., and J. Tobin. 1972. "Is Growth Obsolete?" In *Economic Growth.* NBER Fiftieth Anniversary Colloquium.

Robinson, J. 1972. "The Second Crisis of Economic Theory." *American Economic Review* 62: 1–9.

Russell, C. S. 1973. *Residuals Management in Industry: A Case Study of Petroleum Refining.* Baltimore: The Johns Hopkins University Press for Resources for the Future, Inc.

Smith, V. K. 1972. "The Implications of Common Property Resources for Technical Change." *European Economic Review* 3: 469–479.

Library of Congress Cataloging in Publication Data

Smith, Vincent Kerry, 1945–
Technical change, relative prices, and environmental resource evaluation.

Includes bibliographies.
1. Environmental policy—Mathematical models.
2. Externalities (Economics)—Mathematical models.
3. Technological innovations—Mathematical models.
I. Title.
HC79.E5S54 333.7′01′5118 74-6840
ISBN 0-8018-1626-2